Fig Trees Galore: The Global Encyclopedia

51. Religious festivals and the symbolism of the fig tree

52. Fig trees in the art of calligraphy

53. The fig in children's stories and legends

54. Fig as a source of antioxidants and nutrients

55. The use of fig roots in traditional medicine

56. Fig trees in royal and imperial gardens

57. Fig in natural homemade cosmetics

58. The fig in Asian culinary culture

59. Fig trees in historical travelogues

60. The fig and viticulture: the art of fig wine

61. Fig trees in wedding and birth traditions

62. The influence of the fig tree on architecture and design

63. Dried figs: history, preparation and use

64. The fig and indigenous spirituality

65. The use of fig leaves in food

66. Fig trees in Zen gardens and meditative spaces

67. Figs and alternative medicine practices

68. The fig in poetry and folk song

69. Fig trees in the legends of indigenous peoples

70. The fig and contemporary molecular cuisine

71. The use of fig roots in crafts

72. Fig trees in Middle Eastern culture

73. The fig and food sustainability

74. Fig trees and Chinese healing practices

75. The art of making fig jam

76. The fig in esoteric beliefs

77. Fig trees in traditional African culture

Chapter 1: Botany and classification of the fig tree

Botany and Classification of the Fig Tree

The fig tree, scientifically known as Ficus carica, is a fascinating plant that has captivated the attention of botanists, horticulturists and nature enthusiasts for centuries. Belonging to the Moraceae family, the fig tree is a species distinguished by its distinctive morphology, its history rich culture and its contribution to the surrounding ecosystem.

Botanical Classification

The fig tree belongs to the genus Ficus, which includes more than 800 different species. The botanical classification of the fig tree is as follows:

- **Kingdom:**Plantae
- **Division:**Angiosperms (flowering plants)
- **Class :**Eudicots
- **Order:**Rosales
- **Family :**Moraceae
- **Gender :**Ficus
- **Species:**Ficus carica

The fig tree is a woody, deciduous and relatively fast-growing plant. It is characterized by palmate, lobed and toothed leaves, as well as flowers not visible to the naked eye. The flowers develop inside a structure called a sycone, which is actually a bulging receptacle that uniquely encompasses the flower, fruit and seeds.

Morphology and Characteristics

The fig tree appears in the form of a tree or shrub, generally of medium size. Its leaves measure between 4 and 25 cm long and are deeply lobed, giving them a characteristic hand-shaped appearance. The edges of the lobes may be smooth or slightly toothed, and the color of the leaves varies from dark green to lighter green depending on the variety.

The fruit of the fig tree, of course, is the fig. This unique fruit is actually a fleshy receptacle resulting from the expansion of the syconium. Fig is a multiple type fruit, which means that it contains many small flowers inside. It comes in a variety of shapes, colors and sizes, ranging from green to purple, brown and black.

Geographic Distribution and History

Native to the Mediterranean region, the fig tree has a long history of cultivation dating back to ancient times. Fig trees were grown in regions from ancient Egypt to Greece and Rome, and they played an important role in the cultures and religious rituals of these civilizations. Over time, the fig tree spread across the world through human migration and trade, adapting to a variety of climates and soils.

The botany and classification of the fig tree reveal a plant that embodies the diversity and complexity of the plant kingdom. From the Ficus genus to the Moraceae family, including the unique morphological characteristics of its leaves and fruits, the fig tree represents a harmonious fusion between nature and human culture. Its rich cultural history, ecological properties and contribution to human nutrition make it a fascinating subject of study for botanists and nature lovers.

Chapter 2: The Symbolism of the Fig Tree Throughout History

The fig tree, Ficus carica, has always been much more than just a plant in the history of humanity. It has established itself as a powerful symbol, carrying within it profound meanings that transcend cultures and eras. From ancient myths to religious traditions, the fig tree has captured the imagination and expressed universal concepts through its leaves, fruits and iconic silhouette.

A Sacred Presence in Myths and Legends

The fig tree has often been linked to mythological stories, giving an aura of mystery to its symbolism. In Greek mythology, the fig tree is associated with Dionysus, god of wine and fertility. According to legend, Dionysus was born under a fig tree, and the tree was considered sacred in his honor. Likewise, in the Bible, the fig tree is present in significant stories, such as that of the curse of the barren fig tree by Jesus, a symbol of uselessness and spiritual sterility.

The Fig as Cultural Metaphor

The fruit of the fig tree, the fig, is a powerful metaphor used in many cultures to illustrate complex ideas. In Buddhism, the fig is sometimes used to symbolize the illusion of the material world. Due to its ephemeral nature and soft texture masking a seed-laden interior, the fig can conjure up the idea of deceptive outward appearance.

The Fig Tree in Religions and Spiritual Beliefs

The fig tree plays a significant role in religions and spiritual beliefs across the world. In Islam, it is said that the Prophet Muhammad had a fig tree under which he meditated, thus connecting him to contemplation and spiritual connection. The fig tree is also mentioned in Jewish tradition, symbolizing fertility and prosperity.

An Analogy of Growth and Transformation

The fig tree, with its cycle of growth, fruiting and leaf fall, can be interpreted as a metaphor for human life. Its passage through the seasons can be compared to the different phases of existence, from birth to death, through growth and maturity.

The symbolism of the fig tree throughout history is a remarkable illustration of how plants can transcend their status as simple living organisms to become cultural and spiritual symbols. From the mysticism of ancient myths to the wisdom of religions and beliefs, the fig tree has left an indelible imprint on the way humanity perceives the natural world and expresses intangible concepts.

Chapter 3: The Quest for the Perfect Fig Variety

For millennia, humanity has engaged in a perpetual quest to improve fig varieties, tirelessly searching for the perfect fig. This fascinating quest combines art, science and tradition, and testifies to the fascination that figs exert on the palates and minds of individuals throughout time. The quest for the perfect fig variety is an exploration that transcends the simple cult of flavor and immerses itself in the richness of botanical, cultural and gastronomic diversity.

The Legacy of Selection and Cultivation

Since the earliest days of agricultural cultivation, human beings have noticed that certain figs have superior qualities compared to others. Wild figs were the precursors of this exploration, offering clues about what characteristics to look for to achieve the perfect fig. Early growers began selecting specimens that had the mildest taste, most pleasant texture, and best adaptability to their environment.

The Art of Hybridization and Selection

With the advent of systematic agriculture, the art of hybridization and selection of fig varieties developed. Horticulturists began crossing different varieties to combine the best characteristics of each species, seeking to improve the sweetness, size, color and texture of figs. This quest required both scientific knowledge and intuition cultivated over generations.

Culture and Transmission of Knowledge

The quest for the perfect fig variety was also shaped by the oral and written transmission of knowledge. Farming communities have shared their observations, growing techniques, and breeding secrets across generations. Stories of success and failure have enriched the collective understanding of fig cultivation and inspired new explorations.

Cultural and Gastronomic Diversity

The quest for the perfect fig variety has not been limited to geographic boundaries. Each culture has brought its own approach to the improvement of figs, resulting in an astonishing diversity of varieties across the world. From the velvety purple figs of Provence to the golden figs of the Middle East, each variety reflects the culinary preferences and unique terroirs of its region of origin.

The Infinite Quest

Yet, despite millennia of selection and crossbreeding, the quest for the perfect fig variety remains unfinished. Each new discovery, each new variation pushes the limits of the taste experience and nourishes the collective imagination. Nature itself continues to offer surprises, and future generations will always have new varieties to explore and enjoy.

The quest for the perfect fig variety is a testament to the deep and complex relationship between human beings and the natural world. It reveals our desire for exploration, innovation and connection with the land that nourishes us. This quest continues to invite us to celebrate the diversity of figs and savor the results of our perseverance and creativity.

Chapter 4: The First Historical Traces of the Fig Tree

The fig tree, Ficus carica, has a deep-rooted history dating back to the dawn of human civilization. Through the centuries, it has witnessed the evolution of society, commerce, culture and religion, leaving behind fascinating historical traces that reveal its central role in the development of society. humanity.

Distant Origins and Domestication

The first historical records of the fig tree date back to around 9400 BC. BC, during the Neolithic period, in the Fertile Crescent region, which stretched from Mesopotamia to ancient Egypt. Wild fig trees were first domesticated in these regions, marking the beginning of a close relationship between humans and this valuable plant. Wild figs were an essential food source for early farming communities.

Figs in Antiquity

The fig tree was revered in ancient cultures, playing a major role in myths and legends. In Greek mythology, fig trees were associated with deities such as Dionysus and Hermes. The Greeks and the Romans also adopted the fig into their diet, and figs were often given as offerings to the gods.

Figs in Religious Writings

Fig trees have a significant place in religious texts. In the Bible, fig trees are mentioned several times, symbolizing knowledge (the story of Adam and Eve) and fertility. Fig trees also feature in Islam, notably in the writings of the Prophet Muhammad who praised them for their abundance and usefulness.

Figs in Commerce and Diplomacy

Over time, figs gained considerable commercial importance. Dried figs were a valuable source of food and were often traded along trade routes. The story goes that Egypt's Queen Cleopatra used figs to negotiate access to water for her kingdom with Roman leader Mark Antony.

Figs and Social Symbolism

Figs were also linked to social norms and customs. In ancient Greece, figs were a favored gift for hosts and were often served at banquets. Figs were also considered a symbol of wealth and prosperity.

The first historical traces of the fig tree evoke a rich and complex history that transcends geographical borders and eras. From its humble origins in the Fertile Crescent to its spread across the globe, the fig tree has woven its own story within human history. Figs have nourished the bodies and minds of the ancients and continue to be a living link between our past and our present, illustrating how a simple plant can leave an indelible mark on culture, religion and society.

Chapter 5: The Global Migration of the Fig Tree: A Botanical and Cultural Journey

The fig tree, Ficus carica, has undertaken an extraordinary journey across continents and centuries, becoming one of the most universally loved and cultivated fruit trees. The global migration of the fig tree is both a story of botanical diffusion and a testimony to the cultural and gastronomic influence that this plant has had on different human societies.

Mediterranean Origins

The history of the fig tree begins in the Mediterranean region, where wild fig trees evolved and were domesticated thousands of years ago. Ancient civilizations, from the Greeks and Romans to the Egyptians, grew fig trees for their sweet fruits and versatile leaves. Figs were not only a food source, but also a medicinal resource, as well as a cultural and religious symbol.

Towards New Horizons

À Throughout history, figs have traveled far beyond their land of origin. Population movements, trade and exploration helped spread fig trees across Eurasia, Africa and beyond. Travelers and traders carried cuttings of fig trees, ensuring their spread to new territories. Empires were shaped in part by the presence of thriving fig crops, contributing to the rise of entire regions.

Diversified Climates

The fig tree has proven its ability to adapt to a variety of climates. From the heat of the Mediterranean to the mildness of the Middle East, to the tropics of Asia, fig trees have found ecological niches in varied regions. This adaptability has allowed this species to become a common element of the landscapes and cultures of many countries.

Cultural and Gastronomic Influence

The global migration of the fig tree has not only shaped ecosystems, but also left a indelible mark on local cultures and cuisine. Figs have been incorporated into traditional dishes, from exquisite desserts to savory dishes. They have become symbols of prosperity, generosity and hospitality in many cultures.

The Exchange of Knowledge

The migration of the fig tree also led to an exchange of botanical and agricultural knowledge. Methods for growing, pruning and preserving fig trees were shared and adapted to local climates, strengthening sustainable agricultural practices and food security.

The global migration of the fig tree is much more than just a geographic spread. It is a living story of adaptation, cultural exchange and human connection with nature. By crossing borders and integrating into a myriad of societies, the fig tree has transcended its status as a plant to become an ambassador of the world's botanical, cultural and gastronomic diversity.

Chapter 6: Propagation and Multiplication of Fig Trees: An Ancient Art and a Modern Science

The propagation and multiplication of fig trees constitutes a fascinating field which combines tradition and innovation, ancient know-how and modern discoveries. These processes have evolved over time, reflecting man's passion for this precious plant and his constant quest for improved varieties. From the rooting of ancestral methods to the integration of scientific advances, the propagation and multiplication of fig trees illustrate the interconnection between cultural heritage and botanical research.

Traditional Methods of Multiplication

Since ancient times, vegetative propagation techniques have been used to propagate fig trees. Among the traditional methods, cuttings have been one of the most common. Fig tree cuttings, taken from healthy branches, are rooted in suitable soils. Experienced farmers had a keen eye for choosing promising cuttings, thus ensuring the continuity of popular varieties.

The Art of Grafting

Grafting is another traditional method of propagating fig trees, which has been refined over the centuries. Bud grafting, where a bud is implanted onto a rootstock, has made it possible to retain the desired characteristics of existing varieties while promoting growth. Master grafters were able to create trees from several varieties grafted onto the same trunk.

Scientific Innovation

With advances in science, new methods of propagating fig trees have emerged. Micropropagation, a technique for growing plant tissue in the laboratory, has enabled the mass production of genetically identical fig plants. This method offers a faster alternative to traditional propagation, especially for rare or difficult-to-propagate varieties.

Hybridization for New Varieties

Hybridization also played a crucial role in the propagation of fig trees. By crossing different varieties of fig trees, breeders can create new varieties with improved characteristics, such as higher yields, disease resistance or improved taste qualities. This combination of ancient breeding methods and modern techniques has further diversified the options available.

Preservation of Biodiversity

The propagation and multiplication of fig trees also play an essential role in the preservation of plant biodiversity. While some varieties of fig trees are threatened by habitat loss or climate change, propagation efforts help preserve and share these genetic treasures for future generations.

The propagation and multiplication of fig trees is a harmonious fusion of tradition and modernity. By combining ancient knowledge with contemporary technology, botanists, farmers and fig enthusiasts have managed to preserve and improve this iconic plant. This process reflects the symbiosis between human culture and the plant kingdom, where the diversity of fig trees is carefully maintained to continue to amaze the senses and nourish the bodies.

Chapter 7: Figs in Art and Literature: A Sensory and Symbolic Exploration

Figs, these sweet, fleshy fruits, have long captivated the imagination of artists and writers. Their sensual shape, rich colors and captivating flavor have made them a popular subject in art and literature throughout the ages. Beyond their visual and taste beauty, figs have also been charged with symbolism, evoking themes ranging from sensuality to abundance, knowledge and transformation.

Figures of Art

Figs have often been depicted in works of art, whether paintings, sculptures or photographs. Their distinct appearance, with their oval shape and exposed fleshy interior, has inspired artists to capture their essence in a variety of artistic styles. From sumptuous still lifes to more symbolic depictions, figs have become iconic subjects that reflect the sensual nature of life.

Literary Figuration

Figs have also found their place in literature, as complex and evocative symbols. In poetry, they are sometimes used to evoke passion, lust and sensory experience. Descriptions of their sweetness, juiciness and texture became metaphors for deep emotions and intimate relationships.

Symbols and Metaphors

Figs have been endowed with rich and varied symbolic meanings. In some cultures, they represent abundance and fertility, evoking the generosity of nature. In others, they are linked to knowledge, perhaps because of their association with the biblical story of Adam and Eve. Figs can also symbolize transformation, going from a seemingly unassuming fruit to a source of sweet delight.

The Power of Evocation

The use of figs in art and literature shows these fruits' ability to evoke a complex range of emotions and ideas. The simple act of biting into a juicy fig can evoke a multitude of sensations and memories. This powerful evocation has inspired authors and artists to incorporate them into their creations, creating works that provoke emotional and intellectual responses.

Figs, with their beauty, flavor and symbolism, have woven a golden thread through art and literature. They remind us that simple things can have deep and multiple meanings, whether in the visual representation of a still life or in the poetic metaphor of a human emotion. Figs continue to inspire, nourish and move, making their presence in art and literature a celebration of life itself.

Chapter 8: The Detailed Anatomy of a Fig: A Hidden World of Shapes and Flavors

The fig, this sweet and fleshy fruit, hides within itself a complex anatomy which reveals a symphony of textures, colors and flavors. With just a glance, one might underestimate the depth of this structure, but delving into its anatomy reveals a world of botanical complexity and sensory experiences.

Soft Exterior and Distinguished Skin

The fig starts with its exterior, a smooth and often velvety skin that protects this sweet treasure. Color varies depending on variety, ranging from green to purple to brown to black. The skin not only looks aesthetically pleasing, but it also acts as a protective barrier against parasites and dehydration.

The Robust and Fleshy Receptacle

Slicing a fig in half reveals its fascinating interior. The fruit is actually a fleshy receptacle called a syconium. This unique bulge envelops both flowers, seeds and edible tissues. This is where the magic of fruit ripening and transformation takes place.

The Central Cavity and the Small Flowers

The central cavity of the syconium houses the small flowers. These flowers are not visible to the naked eye, but they are essential for the reproduction of the plant. This is where the pollination process takes place, where tiny insects can play a crucial role in fertilization.

Succulent Pulp and Sweet Nectar

The fleshy pulp surrounding the small flowers is what we know as the edible part of the fig. It can vary in color and flavor depending on the variety. The texture ranges from tender to chewy, and the flavor is a complex combination of sweetness and sometimes slightly tart notes. This pulp is also rich in nutrients and fiber.

Small Crunchy Seeds

Digging a little deeper into the fig, we discover small crunchy seeds scattered throughout the pulp. These seeds are not only edible, but they also add an interesting texture to the experience of eating a fig. They are usually small and often overlooked, but they are an essential part of the fig's anatomy.

The Magic of Taste and Experience

The anatomy of a fig reveals a symphony of textures and flavors that combine into a unique sensory experience. The different layers, from the hidden petals to the crunchy seeds, combine to create this inimitable taste that varies from one variety to another. The act of eating a fig becomes a taste and tactile exploration, a connection with nature and an appreciation of its diversity.

The anatomy of a fig is much more than just a botanical structure. It is a work of nature art, a complex composition of shapes, colors and flavors that intrigues the senses and evokes a deep appreciation for the diversity and beauty of the plant world. Every time we taste a fig, we witness this fascinating anatomy and engage in an experience that transcends the limits of science to touch the heart of the human experience.

Chapter 9: The Different Species of Wild Fig Trees: An Astonishing Botanical Diversity

Wild fig trees, belonging to the genus Ficus, are a diverse family of trees and shrubs that inhabit varied ecosystems around the world. For millennia, these species have coexisted with nature, playing essential roles in ecosystems and influencing human cultures. From Africa to Asia, from the Americas to Oceania, the different species of wild fig trees are a source of botanical wonder and a testament to the ingenuity of plant life .

Extended Biodiversity

The Ficus genus is vast, bringing together more than 800 different species. Of these, some are small trailing plants, while others develop into majestic trees. Each species has its own distinctive characteristics, adapting to the planet's diverse climates and habitats.

Ficus Carica: The Domesticated Fig Tree

Ficus carica, the commonly cultivated fig tree, is one of the best-known species of the genus. Native to the Mediterranean region, it is widely cultivated for its delicious sweet fruits. This species has played an important historical and cultural role, being mentioned in ancient texts and being an integral part of the cuisine and rituals of different cultures.

Ficus Benghalensis: The Giant Banyan

Ficus benghalensis, also called the banyan tree or giant banyan tree, is an impressive species that is revered in many cultures. Originally from India, it is known for its aerial growth mode, where its aerial roots descend from the trunk and take root in the soil to form a complex network. These massive trees are often considered sacred and have great spiritual significance.

Ficus Elastica: The Rubber Tree

The species Ficus elastica, or rubber tree, is native to tropical Asia. It is valued for its latex, which has historically been used to make rubber. Additionally, its large, glossy leaves make it a popular houseplant in areas where the climate does not permit outdoor growth.

Ecological Interactions and Sustainability

Wild fig trees have a unique place in ecosystems because of their symbiotic relationship with specific pollinating insects, called fig trees. Fig trees and wild fig trees depend on each other for survival and reproduction. This interaction demonstrates how nature has evolved to create complex connections between species.

The different species of wild fig trees illustrate the extraordinary diversity of the plant world and how plants have evolved to occupy specific ecological niches. From their roles in ecosystems to their cultural contributions, wild fig trees offer a fascinating insight into the harmonious coexistence between nature and humanity. These species deserve our attention and preservation because they are a living testimony to the ingenuity of life on Earth.

Chapter 10: The Fig Tree and Surrounding Ecosystems: A Pillar of Biodiversity and Balance

The fig tree, Ficus carica, is not simply a fruit tree, but a vital player within surrounding ecosystems. Its impact on biodiversity, ecological regulation and environmental sustainability is profound. As a species revered by humans and harmoniously integrated into nature, the fig tree plays an essential role in the preservation of life and the balance of ecosystems.

Close Connection with Wildlife

Fig trees are renowned for their crucial role in maintaining biodiversity. Their fleshy, sweet fruits are a food source for a variety of animals, such as birds, mammals and insects. By attracting these species, fig trees contribute to cross-pollination, thereby promoting the genetic diversity of plants in their environment.

Shelter and Refuge

Fig trees also provide essential shelter and refuge for many creatures. The tree's dense foliage provides cover from the elements and predators. Small animals can find shelter among the branches, while birds can build their nests safely in the recesses of the tree.

Integrated Life Cycle

The fig tree is also a key player in the recycling of nutrients in ecosystems. Fallen leaves and decomposed fruits enrich the soil with organic matter, which nourishes other plants and creates an integrated life cycle. This contribution to nutrient cycling helps maintain the overall health of the ecosystem.

Role of Ecological Regulation

Fig trees have a regulatory role in ecosystems, helping to control the population of animals and plants. For example, the presence of fig trees can influence the distribution of insect populations by providing habitat for natural predators. Additionally, by providing a food source for a variety of animals, fig trees help keep food chains balanced.

Culture and Nature in Harmony

The interaction between the fig tree and surrounding ecosystems illustrates how nature and culture can coexist harmoniously. Fig trees have been enjoyed and cultivated by humans for millennia, but they have also followed their own ecological path, interacting with other species to support the entire ecosystem.

The fig tree is not only a provider of sweet delights, but also a pillar of life in ecosystems. From providing food and shelter to ecological regulation and preserving biodiversity, the fig tree demonstrates how plants can shape and support the nature around them. By understanding and preserving the vital role of fig trees in ecosystems, we contribute to the health and sustainability of our global environment.

Chapter 11: The Pollination of Fig Trees and the Birth of Figs: A Natural Ballet of Life and **Transformation**

The birth of figs is a marvel of nature that arises from a complex and intimate process of pollination. Fig trees, belonging to the genus Ficus, evolved with a unique symbiotic relationship with specific insects, creating a natural ballet of life and transformation that ultimately results in the creation of figs. The pollination of fig trees is a telling example of how nature orchestrates delicate interactions to ensure the reproduction and survival of plant species.

The Essential Role of Fig Trees

Pollination of fig trees is primarily dependent on specific tiny insects, such as fig trees or fig wasps. These insects have a close relationship with fig trees, as they depend on figs for reproduction and food. In return, fig trees rely on these insects to ensure their pollination. This is a powerful example of co-evolution, where the two parts have developed mutual dependence over millions of years.

The Dance of Pollination

The process of pollination of fig trees begins when female wasps seek out mature figs to lay their eggs. As they enter the fig, they cover themselves with pollen from the male flower, which they previously visited. During their stay in the fig, the wasps lay their eggs and pollinate female flowers by dispersing the pollen they carry on their bodies.

The Transformation of Flowers into Fruits

Once the figs have been pollinated, the female flowers begin to turn into fruit. Wasps that have laid their eggs generally do not survive in mature figs because they do not have enough resources to feed and grow. However, the wasps' pollination and egg-laying processes triggered the growth of the figs and the development of the seeds inside.

The Ripening and Birth of Figs

Over time, figs ripen and transform into the sweet delights we know. The seeds inside have also matured, ready to be dispersed and germinate if they find suitable conditions. Figs thus provide a feast for a variety of animals and ensure the dispersal of seeds to new locations, helping to spread fig trees.

Reflections on Natural Coexistence

The pollination of fig trees and the birth of figs are a living testimony to the harmonious coexistence between plants and insects. This natural ballet, which takes place in the secret of each fig, recalls the subtle and sometimes surprising interconnection that maintains the balance of ecosystems. It also shows us how nature has developed ingenious solutions to ensure the reproduction of plant species, creating a world of beauty and sustainability.

The pollination of fig trees and the birth of figs is a powerful reminder of the complexity and beauty of life on Earth. This intimate process of reproduction and transformation, gracefully orchestrated by nature, invites us to contemplate the invisible dance that takes place in each fig we taste. It is a humbling reminder of the magic that lies in natural interactions and the vital role each species plays in preserving life.

Chapter 11: The Pollination of Fig Trees and the Birth of Figs: A Tale of Symbiosis and

Fertility

In the tranquil landscapes where fig trees thrive, an extraordinary natural spectacle unfolds silently: the pollination of fig trees and the birth of figs. It's a tale of close symbiosis between fig trees and pollinating insects, a complex dance that results in the creation of these fleshy, sweet fruits that have been enjoyed by humans for millennia. This process, both biological and poetic, illustrates how nature creates abundance through the harmonious interaction between plants and creatures.

Intimate Symbiosis

Fig pollination relies on an intimate relationship between fig trees and specific insects, such as fig trees and fig wasps. Fig trees depend on these insects for reproduction, while insects depend on figs for reproduction. Figs, in fact, are transformed inflorescences inside which tiny flowers and insects live in harmony, creating a balanced symbiosis.

The Pollination Process

The pollination process begins when male fig trees produce flowers that contain pollen. Male fig wasps emerge from these flowers and leave the fig trees to find the female figs that are ripening. While searching for female figs, fig wasps carry pollen, thereby pollinating the female flowers inside the figs.

The Birth of Figs

When female wasps find mature female figs, they go inside to lay their eggs. During this process, the wasps transfer the pollen they have collected into the male figs, allowing pollination of the female flowers. Female figs, once pollinated, begin to develop and mature creating favorable conditions for seeds to form.

Support for Biodiversity

The pollination of fig trees also helps support the biodiversity of the ecosystem. Figs attract various types of animals, such as birds and small mammals, which feed on the fruits. By eating the figs, these animals help disperse the seeds, helping the fig trees colonize new places and maintain their population.

Natural Balance

The pollination of fig trees and the birth of figs is a vivid illustration of the natural balance and complex interactions that underpin life on Earth. This narrative, while seemingly simple, reveals a depth of interdependence that keeps ecosystems in harmony. Fig trees and their pollinators are woven together in a delicate weft that reminds us that beauty and abundance often arise from the subtle cooperation between different life forms.

The pollination of fig trees and the birth of figs is an eloquent testimony to how nature orchestrates complex connections to support life and fertility. This symbiotic dance between fig trees and pollinating insects is a fascinating example of how species interact to ensure their survival and reproduction. Figs, these delicious and sweet fruits, carry within them the secret of a story of collaboration, transformation and perpetuation of life.

Chapter 12: The Growing Seasons of the Fig Tree: A Journey Through the Rhythms of Nature

The fig tree, silent witness to the passage of time, follows a growth cycle that reflects the changing seasons and the pulsations of nature. From winter dormancy to summer bloom, the fig tree's growing seasons provide a captivating window into how plant life adapts and thrives throughout the months. This cycle, marked by distinct stages, illustrates the way in which the fig tree interacts with its environment and the elements that influence its development.

Winter Dormancy: Patient Waiting

During the winter months, the fig tree enters a dormant period. Cooler temperatures and shorter days reduce the tree's metabolic activity. Leaves fall, leaving branches bare and vulnerable. It is a time of rest and recovery for the fig tree, where it conserves its energy for seasons to come.

The Emerging Spring: The Awakening of Life

With the arrival of spring and the lengthening of the days, the fig tree comes out of its dormancy. New soft green leaves begin to grow on the branches, announcing the renewal of life. The buds bloom into magnificent inflorescences, setting the stage for the pollination process. It is a time of anticipation, where the tree prepares to bear the fruits of its labor.

Fruitful Summer: The Blossoming of Figs

Summer is the blooming season for the fig tree. The pollinated flowers transform into young figs which grow and ripen with the generous sun. The leaves provide welcome shade to the developing fruit, and the figs gain size and flavor day by day. It is at this stage that the magic happens, transforming the flowers into fleshy, sweet fruits.

Ripe Autumn: Harvest and Decline

À As fall approaches, figs reach full maturity. It's harvest time, where fruits are carefully picked by hand to be enjoyed fresh or made into various delicacies. Leaves are beginning to show warm hues of orange and red, signs of an impending shift into winter dormancy. Fig trees can produce a second, smaller harvest in fall, providing extended bounty.

Celebration of Cyclical Life

The growth cycle of the fig tree illustrates how nature follows a seasonal rhythm that reflects the balance between rest and activity. Each season has its own meaning in the life of the fig tree, and together they form a story of renewal, growth, fruiting and preparation for winter. The growing seasons of the fig tree remind careful observers of the beauty of cyclical life, where each phase has its role to play in the grand picture of nature.

The growing seasons of the fig tree are an invitation to connect more deeply with the natural rhythm of the Earth. This cycle provides an opportunity to celebrate the emergence of new leaves, the blossoming of figs, and the ongoing transformation that characterizes plant life. As we observe the growing seasons of the fig tree, we witness how nature guides each stage of this journey, from winter sleep to summer splendor and perpetual rebirth.

Chapter 13: The Mysteries of the Fig: Between Legends and Beliefs

The fig, this sweet, fleshy fruit, is shrouded in mysteries that have captivated the imagination of cultures throughout the centuries. Beyond its sweet flavor, the fig has been imbued with legends, beliefs and deep symbolism. From ancient mythology to spiritual significance, the mysteries surrounding the fig add an extra layer of fascination to this humble and delicious fruit.

In the Shadow of Ancient Myths

Figs have often been linked to myths and legends in various cultures. In Greek mythology, fig trees were considered sacred to Dionysus, the god of wine and fertility, and figs were often associated with mystical knowledge and the bounty of nature. In the biblical Creation story, the fig symbolized the understanding of truth and duality, as illustrated in the story of Adam and Eve.

The Symbolism of the Fig Leaf

The fig leaf also has notable symbolic meaning. In many cultures it has been used to represent protection, modesty and covering. In the biblical context, Adam and Eve used fig leaves to cover themselves after becoming aware of their nakedness. This

The symbolism of the fig leaf then expanded to represent modesty and the need to protect oneself.

The Fig in Spiritual Beliefs

In some beliefs, the fig has been associated with spirituality and inner transformation. Its fleshy and juicy form has been interpreted as a symbol of the human soul and its hidden depth. The fig has often become a metaphor for expressing ideas about self-discovery, inner knowing and spiritual journey.

Rituals and Traditional Uses

Figs have also played a role in various traditional rituals and uses. In some cultures, they were given as offerings to the gods for blessings and bountiful harvests. Figs were also used to prepare ointments and potions in traditional medicine, being associated with healing properties and vitality.

Between Mysticism and Reality

The mysteries surrounding the fig have added a mystical dimension to this common fruit. Whether it's its connection to ancient deities, its symbolic roles, or its associations with spirituality, the fig is much more than just a delicious treat. She embodies the hidden depths of human history, mythology and belief, offering insight into how cultures have found profound meanings in the simplest elements of everyday life.

The mysteries of the fig reveal how human beings have found deep meanings in the nature around them. These legends, beliefs and symbolism remind us that fruits are not only sources of nutrition, but also bearers of cultural and spiritual meaning. The fig, with its long history of mystery and meaning, invites us to look beyond its sweetness and discover the enchanting tales that have been woven around this fruit for centuries.

Chapter 14: The Cultural Importance of the Fig Tree in Ancient Societies: An Unwavering Link
between Man and Nature

In the recesses of ancient history, the fig tree has held a place of honor in cultures across the world. Much more than a simple fruit tree, the fig tree has been a witness and actor in stories, myths and beliefs that shaped ancient societies. Its role as a food source, spiritual symbol and cultural element is a captivating illustration of the intimate relationship between man and nature.

Abundant and Vital Food

In ancient societies, the fig tree provided an essential food source. Figs, rich in nutrients and natural sugars, were a source of vitality for populations, serving as a nutritional supplement in the varied diets of the time. Dried figs, easy to store, were a source of provisions for lean seasons, thus guaranteeing food security.

Symbol of Prosperity and Fertility

The fig tree was frequently associated with ideas of prosperity and fertility. In many cultures, its green leaves and fleshy fruits were seen as a sign of growth and abundance. Fig trees in full bloom and laden with fruit were often seen as a symbol of blessing and success, representing nature's ability to support and nourish human life.

Mythology and Religiosity

In the mythology and religiosity of various ancient civilizations, the fig tree has played a central role. In ancient Greece, for example, the fig tree was dedicated to Dionysus, the god of wine and fertility. In the biblical context, the fig tree is mentioned several times, notably in the story of Adam and Eve. Its significance as a spiritual element has enhanced its cultural significance, linking it to the fundamental beliefs and spiritual lives of ancient people.

Crafts and Industry

Fig trees also contributed to the crafts and industry of ancient societies. Fig tree fibers were used to make textiles and ropes, while the leaves were used to create utilitarian and decorative objects. These varied uses of different parts of the tree strengthened the relationship between fig trees and the daily lives of ancients.

A Cultural and Social Pillar

The fig tree has been more than just an element of subsistence in ancient societies. It has witnessed social gatherings under its shade, spiritual rituals around its branches and commercial transactions under the shade of its leaves. Fig trees have become cultural landmarks, a link between generations and customs.

The cultural importance of the fig tree in ancient societies transcends its role as a fruit supplier. It has been woven into the fabric of human life, carrying tales of fertility, beliefs and traditions. This humble fruit has acquired meaning that goes far beyond its sweetness, showing how natural elements can become cultural icons, connecting human beings to their roots and the land that nourished them.

Chapter 15: The Fig as a Source of Nutrition and Health: A Natural Treasure of Benefits

For millennia, the fig has been recognized as a gem of nature, offering not only exquisite flavor but also an abundance of health benefits. As a fruit loaded with nutrients and beneficial compounds, the fig has taken its place in diets and wellness practices across the world. Its nutritional profile rich in essential nutrients makes it much more than just a treat – it's a valuable source of vitality and health.

An Abundance of Essential Nutrients

Fig is a wealth of essential nutrients that contribute to the overall health of the body. It is a source of dietary fiber, vitamins (especially A, C and K), minerals (such as potassium, magnesium, calcium and iron), and antioxidants. These elements act in synergy to support various aspects of health, from cell growth to immune function to blood pressure regulation.

Fiber for Digestion and Satiety

Figs are rich in dietary fiber, which makes them a great ally for healthy and regular digestion. Fiber promotes intestinal health by preventing constipation and promoting intestinal transit. Additionally, they contribute to feelings of fullness, which can help control appetite and maintain a balanced body weight.

Antioxidants for Cellular Protection

Antioxidants found in figs, such as polyphenols, flavonoids and carotenoids, have a crucial role in protecting cells against oxidative damage. They help neutralize free radicals, unstable molecules associated with premature aging and various chronic diseases, such as heart disease and certain types of cancer.

Potassium for Cardiovascular Health

Potassium, an abundant mineral in figs, is essential for maintaining electrolyte balance and regulating blood pressure. Adequate potassium consumption is associated with a decreased risk of cardiovascular disease and hypertension. Figs, high in potassium and low in sodium, are a nutritious option to support heart health.

The Benefits of Dried Figs

Dried figs, with their increased concentration of nutrients, are also an attractive nutritional option. They retain most of the nutritional benefits of fresh figs and can be eaten as energy snacks, baking ingredients or dietary supplements.

The fig embodies much more than a simple delicacy; it is a natural treasure of nutrition and health. Its unique combination of nutrients, fiber, antioxidants and minerals makes it a dietary option valuable for supporting vitality, digestion, cardiovascular health and cellular protection. Generations past and present have enjoyed the many benefits of this delicious fruit, attesting to its value as a valuable natural resource for a healthy and balanced lifestyle.

Chapter 16: The Fig in International Gastronomy: A Taste Journey Across Cultures The fig, a deliciously sweet and fleshy fruit, has captured the hearts of gourmets around the world. From Asia to America, via Europe and Africa,

the fig has entered international gastronomic tables, offering a rich and diverse taste experience. Its versatile use and characteristic taste have made the fig a valuable ingredient in traditional and contemporary cuisines, adding a special touch to a variety of dishes.

Mediterranean: The Cradle of the Fig

The Mediterranean region has a long love affair with the fig. Majestic fig trees line the landscapes of Greece, Italy, Turkey and other countries in this region. Fresh or dried figs are often eaten simply as a dessert, but they can also be made into jams, pastries, and accompaniments for cheeses, creating a symphony of sweet and savory flavors.

Asia: Fusion of Flavors

In Asia, the fig finds its place in a variety of dishes, bringing a sweet and exotic touch. In India, figs are used to make sweet and sour chutneys, while in the Middle East they are often incorporated into meat or rice dishes, balancing the flavor profiles. Dried fig is also a popular ingredient in Iranian cuisine dishes.

Europe: The Figuration of Gluttony

In Europe, the fig is a symbol of gluttony and refinement. In Spain, fresh figs or dried are combined with cheese platters, creating a balance between sweet and savory. In France, the fig is often featured in salads with cheeses, nuts and vinaigrettes.

America: Creative Integration

Although the fig is not native to America, it has found its way into the continent's creative cuisines. In the United States, fresh or dried figs are often added to salads and meat dishes, adding a sweet, textural note. In Brazil, they are used to prepare traditional jams and desserts.

Africa: A Taste of Luxury

In Africa, the fig is often seen as a luxurious and refined ingredient. Dried figs are used to prepare confectionery and sweet dishes, adding a natural richness to local cuisine. In Egypt, for example, figs are sometimes stuffed with dried fruits and nuts to create treats popular during celebrations.

A Contemporary Touch

In contemporary cuisine, the fig continues to inspire chefs and food lovers. It can be made into elegant sauces, reductions for grilled meats, toppings for pizza, or even an ingredient for sophisticated cocktails and desserts.

The fig, with its sweet flavor and fleshy texture, has transcended geographical boundaries to become an essential star of international gastronomy. Its versatile use in sweet and savory dishes, as well as its creative adaptation to various cuisines, make it a valuable and appreciated ingredient. Beyond its health benefits, the fig has a special power: that of bringing together different cultures around a common passion for delicious food and culinary creativity.

Chapter 17: Tasty Traditions: Authentic Recipes Using Figs

Traditional recipes using figs are culinary treasures that have been passed down from generation to generation, enriching palates with authentic and memorable flavors. From Mediterranean simplicity to Asian elegance, these culinary creations demonstrate the versatility of the fig and its essential role in world cuisines. Here is a taste journey through some of the traditional recipes that celebrate the fig in all its splendor.

1. Figures & Prosciutto (Italy)

One of the most iconic combinations in Italian cuisine, fig and prosciutto, blend sweetness and saltiness exquisitely. Fresh figs, wrapped in slices of prosciutto, offer a harmony of flavors that tickle the taste buds. Served as an antipasto or starter, this simple but elegant dish is a tribute to the sophisticated simplicity of Italian cuisine.

2. Figgy Pudding (UK)

Figgy Pudding, a traditional British dessert, is a dense, moist cake made from dried figs. Flavored with warm spices and served with a sweet sauce, this pudding is often associated with Christmas and winter celebrations. It embodies the warm comfort of the season while paying homage to the historic importance of the fig in British cuisine.

3. Fig Jam (Greece)

Fig jam is a Greek specialty appreciated for its simplicity and delicacy. Figs are combined with sugar and sometimes lemon to create a sweet and fragrant jam. It is often enjoyed with Greek yogurt or cheese, adding a sweet and tangy touch to these creamy dishes.

4. Mrouzia (Morocco)

Mrouzia is a Moroccan dish made with lamb, dried figs, almonds and spices. This sweet and savory dish is simmered slowly, allowing the flavors to blend together harmoniously. The fig gives sweetness natural that contrasts with the spices, creating a symphony of complex flavors that are typical of Moroccan cuisine.

5. Figs in Syrup (Greece)

In Greece, figs are often prepared in a delicate syrup. Fresh figs are poached in a sweet syrup flavored with spices like cinnamon and vanilla. This dessert is served with Greek yogurt or ice cream, creating a mix of textures and flavors that evoke the richness of the Mediterranean.

6. Fig Rice Pudding (Türkiye)

Fig Rice Pudding, a Turkish sweet, combines the creamy familiarity of rice pudding with the distinctive flavor of fig. Dried figs are rehydrated in milk during cooking, infusing each bite of this classic sweetness with a fruity, sweet note.

Traditional recipes using figs are a testimony to the richness and diversity of global cuisine. Each dish tells a story, connecting people to the lands where fig trees have thrived for centuries. From starters to desserts, from sweet to savory dishes, the fig offers a range of culinary possibilities that honor its sweetness and its unique character. These traditional recipes, full of flavor and memories, are a reminder that the fig is more than just a fruit – it is an invaluable source of inspiration and delight in international cuisine.

Chapter 18: The Art of Pot Growing Fig Trees: A Small Canvas for Big Beauty

The art of pot growing for fig trees is a fascinating way to capture the majesty and flavor of fig trees in a small space. Transforming a pot into a thriving scene, this practice allows urban and small space gardening enthusiasts to create an impressive display of lush leaves and sweet figs. It's a celebration of horticultural ingenuity and concentrated beauty, where a small pot becomes a frame for a miniaturized tree.

The Selection of the Potted Fig Tree

Choosing the right fig tree for growing in pots is crucial. Dwarf or semi-dwarf varieties are generally the best options because they adapt well to containers and are more manageable in terms of size. It is essential to choose a fig tree that suits your region's climate and potted growing conditions.

The Perfect Container

The choice of pot is just as important as that of the fig tree itself. Terracotta pots are often recommended as they allow for better air circulation and efficient drainage. Make sure the pot is large enough to accommodate root growth and to maintain a balance between the size of the tree and the size of the pot.

The Ideal Substrate

A quality substrate is essential for successful pot growing. A well-draining mix that retains moisture without creating water stagnation problems is recommended. Potting mixes made from compost, perlite and vermiculite are often used to provide optimal growing conditions.

Location and Proper Care

The positioning of the potted fig tree is crucial. Place it in a sunny location, as fig trees like direct sunlight for vigorous growth and optimal fruiting. Potted fig trees tend to have more frequent water needs, so monitor the soil and water regularly to prevent drying out.

Pruning and Prevention

Pruning is an essential part of container growing for fig trees. Since space is limited, it is important to prune dead or non-productive branches to maintain the shape and vigor of the tree. Additionally, regular pruning can help control the size of the tree and prevent problems related to overgrowth.

The Reward of the Harvest

One of the most rewarding parts of growing fig trees in pots is harvesting fresh figs. Figs are usually ready to harvest when they are soft to the touch and slightly wrinkled. Fresh figs harvested from your pot provide an incomparable taste experience, bringing together the care, attention and patience invested in pot growing.

The art of pot growing for fig trees is a manifestation of creativity and love for nature in a small space. It is a tribute to the adaptability of nature and the possibility of creating magnificent scenes in urban and limited environments. Pot growing for fig trees is not only a convenient way to get fresh figs, it is also a way to merge beauty, science and passion to create a verdant corner of contemplation and indulgence in your own space.

Chapter 19: Pruning and Training Fig Trees: Sculpting Growth to Maximize Harvest

Pruning and training fig trees are essential practices to ensure healthy growth, abundant fruiting and a well-balanced aesthetic. Fig trees, although they tend to grow naturally in a bushy manner, respond favorably to judicious pruning which encourages the production of succulent figs and makes them easy to care for. It is a horticultural art which combines knowledge, observation and mastery to obtain satisfactory results.

The Basics of Pruning Fig Trees

Pruning fig trees usually begins by removing dead, damaged or diseased branches. This step promotes the overall health of the tree by eliminating areas of decay or disease risk. Next, pruning aims to create an open, airy structure that allows light and air to penetrate the tree, thus promoting fruiting.

Training Size of Young Trees

Training pruning is especially important for young fig trees. The goal is to guide the tree's growth by creating a strong structure with well-spaced main branches. This helps distribute the fruiting load evenly, encourages fig production throughout the tree and makes picking easier. In general, vase-shaped fig trees, with a central trunk and spreading branches, are recommended.

Pruning Established Fig Trees

For more mature fig trees, pruning is often aimed at managing excess growth and controlling the size of the tree. Branches that cross or rub against each other can be pruned to prevent friction and promote air circulation. Branches that grow toward the interior of the tree can be pruned to open the tree to sunlight.

Annual Maintenance Pruning

Annual maintenance pruning is generally recommended for fig trees. This involves pruning non-fruitful shoots and removing dead or damaged branches. Annual pruning promotes more robust fig production because it focuses the tree's energy on the fruit-bearing branches.

The Balance Between Size and Harvest

A crucial aspect of pruning fig trees is finding the balance between pruning and harvesting. Excessive pruning can reduce the harvest because it limits fruit-bearing branches. On the other hand, a lack of pruning can result in untidy growth, poor light penetration and a less abundant harvest.

Pruning and training fig trees is an art of mastery that offers rewards in yield and tree health. By understanding the specific needs of fig trees, gardeners can shape their growth to maximize production of flavorful and healthy figs. Pruning and training, although demanding, are investments that result in fig trees that not only beautify the landscape, but also offer an abundance of delectable softness.

Chapter 20: The Challenges of Growing Fig Trees in Cold Climates: The Delicate Art of Cultivating Sweetness **in Adversity**

Growing fig trees in a cold climate is a challenge that tests the perseverance and ingenuity of gardeners. While fig trees are often associated with warm, Mediterranean regions, gardening enthusiasts in colder climates strive to overcome obstacles to create soft oases in less forgiving environments. It is an endeavor that requires a deep understanding of the needs of fig trees and creativity in finding solutions adapted to adverse climatic conditions.

Choice of Resistant Varieties

One of the first challenges for gardeners in cold climates is choosing cold-tolerant varieties of fig trees. Some varieties are better adapted to cold temperatures than others. Hardy fig trees are generally preferable, as they have the ability to withstand lower temperatures. The search for varieties adapted to the climate is therefore an essential step in successfully growing fig trees in a cold environment.

Winter Protection

Winter protection is a crucial consideration for fig trees in cold climates. Fig trees are vulnerable to frosts and cold winds, which can damage fragile parts of the tree, including young shoots and buds. Gardeners can use methods such as wrapping trees with insulating materials, mulching the soil to protect the roots, and creating temporary structures to provide shelter from the winter elements.

Growing in Containers and Growing in Greenhouses

In cold climates, container growing and greenhouse growing offer viable solutions for grow fig trees. Containerized fig trees can be moved indoors during cold months, providing protection from frost. Greenhouses, by creating a warmer microclimate, allow fig trees to thrive even in difficult climatic conditions.

Management of Growth and Fruiting

Cold climates can slow the growth and fruiting of fig trees. Careful management of pruning, fertilization and irrigation can help stimulate tree growth and encourage fig production. Regular pruning to remove dead or non-productive branches, along with balanced fertilization, can help maintain the health of the tree and optimize fruit production.

Adaptation and Creativity

Growing fig trees in a cold climate requires a certain amount of adaptation and creativity. Gardeners should be willing to experiment with different approaches to find what works best in their specific environment. Challenges can seem daunting, but they also provide an opportunity to push the boundaries of culture and explore new methods for success.

Growing fig trees in cold climates is a demanding adventure, but it offers unique rewards for persistent gardeners. Despite climatic obstacles, each fig harvested becomes a symbol of success and ingenuity. The challenges of growing in cold climates force gardeners to push the boundaries of tradition and explore new ways to cultivate sweetness in less hospitable environments. It is a demonstration of nature's resilience and human determination to create beauty where it is least expected.

Chapter 21: The Interaction between the Fig Tree and the Bees: A Natural Symbiosis of Fulfillment
Mutual

The interaction between the fig tree and the bees is a perfect illustration of the symbiosis between plants and pollinators. These two players in nature have evolved over millennia to depend on each other, creating a harmonious dance that benefits both parties and the ecosystem as a whole. This complex relationship is a true celebration of the interconnectedness of life on Earth.

The Pollinator Relationship

The fig tree and the bees have developed a close relationship where each benefits from the actions of the other. Fig trees are plants pollinated by specific insects, called agaonids or "fig wasps". The flowers of the fig tree are actually inverted inflorescences in which the tiny female flowers are hidden. Agaonids enter the inflorescences to lay their eggs, and in doing so they carry pollen from one flower to another, allowing cross-pollination and the formation of figs.

The Mutuality of Pollination and Reproduction

For agaonid bees, the relationship is just as vital. Fig trees provide an ideal environment for egg laying and reproduction. Female agaonids enter the inflorescences to lay their eggs, and in the process they transfer the pollen that allows the formation of figs. Agaonid larvae develop inside figs, consuming some of the seeds, preparing them to carry pollen when they emerge.

Enriched Biodiversity

The interaction between the fig tree and bees is not limited only to fig trees and agaonids. A variety of other insects, including domestic and wild bees, as well as other pollinating insects, are also attracted to fig flowers to forage for nectar and pollen. This biodiversity enriches the environment and contributes to the overall health of the ecosystem.

Environmental Balance

The interaction between fig trees and bees has a significant impact on environmental balance. There

Pollination of fig trees by bees promotes fruit production, which is vital for wildlife that depends on figs for food. Additionally, fig trees serve as breeding grounds for agaonids, which themselves are an integral part of the food chain for other creatures.

Preserving Interaction

Preserving the interaction between the fig tree and the bees is essential to maintain the balance of the ecosystem. Degradation of bees' natural habitat and anthropogenic disturbance can disrupt this delicate relationship. Protecting natural habitats, reducing the use of harmful pesticides and raising awareness of the importance of pollinators are essential actions to ensure this beneficial interaction continues.

The interaction between the fig tree and the bees is a poignant example of how nature has forged close bonds between plants and animals to promote mutual survival and ecological balance. The fig trees and the bees dance to the rhythm of a perfect symbiosis, where each contributes to the sustainability of the other. This relationship demonstrates the complex beauty of the interaction between different life forms and reminds us of the need to preserve biodiversity for the well-being of our planet and its inhabitants.

Chapter 22: The Fig in Traditional Medicine: A Natural Treasure for Health and Wellness

be

For millennia, the fig has been revered not only for its delicious flavor, but also for its health-promoting properties in various medicinal traditions around the world. The fig, rich in nutrients and natural compounds, has been used to treat a variety of ailments and disorders, reflecting the wisdom of the ancients in natural and holistic medicine.

Balance and Digestion

In many cultures, the fig has been traditionally associated with digestion and the regulation of the digestive system. The abundant dietary fiber present in figs helps boost the bowel movements and prevent constipation. Dried figs, rich in soluble and insoluble fiber, have often been used to soothe gastrointestinal problems and maintain a healthy digestive system.

The Heart and Circulation

Figs offer benefits for cardiovascular health. They contain considerable amounts of potassium, an essential mineral that plays a role in regulating blood pressure. Antioxidants found in figs, such as polyphenols, may help protect blood vessels and reduce the risk of cardiovascular disease.

Diabetes Management

In some cultures, figs have been used to help maintain blood sugar levels. The fiber found in figs may slow the absorption of sugars, which may benefit people with type 2 diabetes. However, it is important to consult a healthcare professional before taking any significant dietary changes to manage diabetes.

Immune Strengthening

Figs are a source of vitamins and minerals, including vitamin C, which plays a key role in strengthening the immune system. The antioxidants found in figs can help protect cells from damage caused by free radicals, helping to strengthen the body's resistance to infections and disease.

General Well-being

In many traditions, figs have been used as general tonics to improve well-being. Their diverse nutritional profile makes them an ideal food for maintaining energy, vitality and overall health. Figs are also rich in minerals such as calcium, magnesium and iron, which support bone, muscle and blood health.

Warnings and Caution

Although figs offer many health benefits, it is important to remember that their consumption should be included as part of a balanced and varied diet. Figs are naturally high in sugar, so excessive consumption can impact blood sugar levels. In addition, some people may be allergic to figs, so it is advisable to introduce them into the diet gradually.

The fig, a natural treasure rich in nutrients and bioactive compounds, has a well-deserved place in traditional medicine. Cultures around the world have recognized and harnessed the beneficial properties of the fig to support health and well-being. However, it is important to view these traditional uses as complementary to modern medical practices and to seek professional advice if medical concerns arise. The fig, as a delicious fruit with therapeutic virtues, remains a reminder of the wisdom of nature and the harmony between man and plant.

Chapter 23: The Fig in Healing Rituals: The Ancient Power of the Sacred Fruit

Since the dawn of humanity, figs have been much more than just a source of nutrition. They have held a special place in the beliefs and healing practices of many cultures around the world. Their sacred status and unique nutritional properties have made figs an essential element in healing rituals, illustrating the deep-rooted role they have played in the quest for health and well-being.

The Symbolism of the Fig

The fig has often been seen as a symbol of fertility, renewal and healing. Its characteristic shape, fleshy texture and mild taste make it a fruit rich in symbolism, evoking life, growth and vitality. In many cultures, the fig is associated with the goddess of fertility and healing, emphasizing its role in promoting health and regeneration.

Purification and Healing Rituals

In some traditions, figs were used in purification and healing rituals. Fresh or dried figs were often eaten or used to prepare infusions to remove toxins from the body, aid digestion and boost immunity. Figs were considered a source of vitality and strength, helping to restore the body's natural balance.

Amulets and Talismans

Dried figs have also been used as amulets or talismans for protection and healing. Worn around the neck or placed under the pillow, they were believed to have the power to ward off illness and promote restful sleep. The nutritional properties of figs were associated with healing and protective qualities that transcended the physical world.

Spiritual and Emotional Rituals

Figs have been involved in rituals to not only heal the body, but also to soothe the mind and emotions. Meditation and ritual consumption of figs were believed to promote inner peace, mental clarity and emotional balance. Figs were seen as a way to strengthen the connection between body and mind, encouraging holistic healing.

The Modern Approach

Although fig healing rituals have often been shrouded in mystery and spirituality, modern knowledge of nutrition has confirmed the health benefits that the ancients intuitively recognized. Figs are rich in antioxidants, fiber, minerals and vitamins, making them favorable for digestive, cardiovascular and immune health.

The fig in healing rituals embodies the ancient wisdom of humanity, connecting nature with physical and spiritual health. The beliefs surrounding figs in healing practices highlight the ways in which cultures have recognized and honored the healing and regenerative powers of this fruit. The fig remains a reminder of the deep interconnectedness between man and nature, as well as a testimony of faith in the Earth's intrinsic capacity to guide towards healing and well-being.

Chapter 24: Remarkable Fig Trees Around the World: The Botanical Giants of the Earth

Fig trees, with their majestic silhouette and lush leaves, have captivated the human imagination for millennia. Across the world, these remarkable trees have thrived in varied environments, leaving a lasting imprint on the natural and cultural history of their respective regions. Whether revered for their age, their imposing size or their central role in local ecosystems, remarkable fig trees are living witnesses to the power of nature and the symbiosis between plants and their environment.

The Banyan Fig Tree in Kalpavriksha, India

The Banyan fig tree (Ficus benghalensis) is revered in India under the name Kalpavriksha, often translated as "wish tree". One of the most famous examples of this type is the Banyan fig tree of Howrah, Kolkata. With an impressive crown that spans approximately 1.5 hectares, this majestic tree is revered as a symbol of fertility, strength and healing. Local worshipers and visitors come to shelter under its soothing shade, creating an atmosphere of respect and worship.

The Moreton Bay Fig Tree, Australia

The Moreton Bay fig tree (Ficus macrophylla) is an iconic species in Australia. One of the most famous specimens is the Curtain Fig Tree, located in Byron Bay. This botanical giant, also known as the "Strangler Fig" (strangler fig), wraps itself around the growing host tree, gently choking the underlying tree over time. Despite its sinister name, the Moreton Bay fig tree is a vital part of the Australian ecosystem, providing shelter and food for many animal species.

The Pagoda Fig Tree, Cambodia

The pagoda fig (Ficus religiosa), also called the "Bodhi Tree", occupies a central place in Buddhist spirituality. The tree is believed to be the place where Buddha achieved enlightenment. The fig tree

Pagodas is a symbol of knowledge, wisdom and spiritual enlightenment. The remarkable tree located at Wat Mahathat in Ayutthaya, Thailand, is famous for its root that grew around a Buddha head, creating an iconic image that embodies the deep connection between nature and spirituality.

The Cursed Fig Tree, Madagascar

The cursed fig tree (Ficus trichopoda) of Madagascar is a stunning example of how nature can be transformed into a sculptural work of art. The massive aerial roots of this tree intertwine and extend dramatically, creating a structure almost sculpted by time. This unique fig tree is a striking example of nature's ingenuity to adapt and thrive in harsh environments.

Remarkable fig trees around the world tell stories of perseverance, symbiosis and respect for nature. These botanical giants embody the majesty and complexity of plant life, while leaving an indelible mark on the crops and ecosystems that surround them. Their presence is a reminder of the power of nature to shape astonishing landscapes and create deep connections between humans and the natural world around them.

Chapter 25: The Enchanted Myths and Legends of the Fig Tree: The Sacred Tree of the Human Imagination

The fig tree, with its majestic stature and abundant leaves, has always been a central character in the myths and legends of cultures around the world. From fertility to spirituality, healing and creation, the fig tree embodies a rich web of symbols and meanings that have captured the human imagination for millennia.

The Fig Tree and Creation

In some cultures, the fig tree is considered the tree of creation, the source of life itself. In Greek and Roman mythologies, the fig tree is associated with the goddess Dionysus (Bacchus), god of wine and fertility. The fig tree is also revered in Hindu tradition as the sacred tree of the goddess

Sarasvati, associated with knowledge, music and art.

The Fig Tree and Spirituality

The fig tree is often associated with spiritual and religious notions. In Buddhist tradition, the pagoda fig (Ficus religiosa) is considered sacred because it was under this tree that Buddha achieved enlightenment. This fig tree, also known as the "Bodhi Tree", symbolizes the quest for wisdom and spiritual truth.

The Fig Tree and Fertility

Due to its propensity to produce a large quantity of fruit, the fig tree is often associated with fertility and abundance. In ancient Greek myths, the fig tree was linked to the fertility goddess, Demeter, and her daughter Persephone. Figs were also given as offerings to fertility deities in many cultures.

The Fig Tree and the Transformations

In mythologies, the fig tree is sometimes linked to magical and mysterious transformations. Strangler fig trees, which wrap their roots around host trees, are often surrounded by legends of transformation and bewitchment. They embody the concept of life arising from death, symbolizing regeneration and the eternal cycle.

The Fig Tree and the Sacred Places

Many ancient fig trees are considered sacred trees, often associated with places of worship and religious sites. These majestic trees, with their imposing presence and eternal character, add a spiritual dimension to the sites where they grow. They become centers of gathering and worship, providing a tangible connection between the divine and the earthly.

The myths and legends linked to the fig tree are proof of the profound impact this tree has had on human imagination. Each culture has woven its own narrative around this majestic tree, reflecting the values, beliefs and hopes of society. The fig tree transcends geographic and temporal boundaries, uniting human beings through a shared fascination with the mysteries of nature and the deeper meanings that emerge from its abundant shade.

Chapter 26: Secrets of Preserving Fresh Figs: Preserving the Sweetness of Nature

Fresh figs, with their sweet flesh and delectable texture, are a seasonal delicacy enjoyed by many fruit lovers. However, their short lifespan can make their conservation a challenge. Discovering the secrets to preserving the flavor and quality of fresh figs can extend the pleasure of tasting them beyond the season.

Choosing the Right Figs

The first step to storing fresh figs is to choose ripe but firm fruits. Avoid figs that are too soft or blemished, as they can go bad quickly. Look for figs that are slightly soft to the touch, but not too soft, with an even, vibrant color.

Refrigerate Immediately

As soon as you bring figs home, place them in the refrigerator. Fresh figs are sensitive to heat and humidity, which can accelerate their ripening and cause them to deteriorate quickly. Place them in a perforated plastic bag or plastic box with a paper towel at the bottom to absorb moisture.

Avoid Premature Washing

It is best not to wash figs before refrigerating them. Excessive humidity can promote mold and deterioration. Wash the figs just before eating them to maintain their freshness.

Consume quickly

Fresh figs tend to go bad quite quickly, even when refrigerated. It is therefore advisable to consume them within two to three days following their purchase. The sooner you eat them, the more you can enjoy their sweet flavor and delicious texture.

Freezing Figs

If you have a surplus of fresh figs and want to keep them longer, you can freeze them. Wash them, remove the stems and cut them into pieces if you wish. Place the pieces on a baking tray and freeze until firm. Then transfer the figs to freezer bags or airtight containers and return them to the freezer. They can be used in smoothies, baked goods and other preparations once thawed.

Creative Use

If you have fresh figs that are starting to ripen, but you can't consume them quickly, consider using them in recipes. Figs can be made into jams, compotes or sauces to extend their life while adding a sweet touch to your dishes.

Fresh figs are a treasure of nature to be enjoyed in season. By understanding the secrets of their preservation, you can maximize their freshness and extend their delicacy to enjoy them beyond their brief period of availability. By following the recommendations for refrigeration, rapid consumption and freezing, you will be able to preserve the natural sweetness of fresh figs and savor each bite with satisfaction.

Chapter 27: Treasures Derived from the Fig Tree: Oils, Lotions and More

The fig tree, in addition to offering succulent fruits, presents a hidden richness in its derivatives. From oil to lotions, including beauty and wellness products, products from the fig tree have captivated the attention of consumers.

connoisseurs looking for natural care. These treasures derived from the fig tree have the power to bring benefits of this exceptional plant in new forms and new experiences.

Fig Essential Oils

Fig essential oils are increasingly popular for their beneficial properties for the skin and health. Rich in antioxidants and fatty acids, these oils can hydrate and nourish the skin, leaving it soft and supple. They can also be used to create unique and calming scents. Fig essential oils lend themselves to aromatherapy, offering refreshing scents that can promote relaxation and well-being.

Fig-Based Lotions and Creams

Fig-based lotions and creams have become popular in the skin care industry. Grace
à Their hydrating and soothing properties, these products can help maintain the elasticity and health of the skin. The vitamins, minerals and antioxidants found in figs help to deeply nourish the skin and protect against environmental damage.

Natural Beauty Products

Figs are increasingly being incorporated into natural beauty products, such as facial masks, scrubs and serums. Due to their content of essential nutrients, figs can revitalize the skin, promote a glowing complexion and reduce the signs of aging. Additionally, their natural softness makes them suitable for sensitive skin.

Derived Food Products

In addition to skincare and beauty, figs are also used in a range of food-derived products. Products such as jams, fig vinegars, teas and infusions add a sweet and delicious note to various culinary preparations. Figs are also a source of fiber and nutrients, making them a great choice for healthy and flavorful food products.

The appeal of nature

The growing popularity of products derived from the fig tree reflects the growing trend toward natural skincare and authentic ingredients. Consumers are looking for natural alternatives to chemicals and synthetics, and products derived from the fig tree offer a solution that combines the power of nature with modern innovation.

Products derived from the fig tree embody the diversity of benefits that this extraordinary plant offers to humanity. Whether used for beauty, health or gastronomy, these derivative treasures captivate attention by providing sensory experiences and holistic benefits. They are a testament to the power of nature to provide versatile and effective resources that meet the needs of our body, mind and well-being.

Chapter 28: The Fig: An Ancestral Religious and Spiritual Symbol The fig, with its rich nuances of religious and spiritual symbolism, has marked the beliefs and spiritual practices of various cultures throughout the ages. As an ancient tree and nourishing fruit, the fig has become a powerful symbol that evokes notions of knowledge, spirituality, regeneration and deep connection between man and the divine.

The Fig in the Scriptures

Figs play a significant role in many religious traditions. In biblical texts, the fig is mentioned several times. In the Old Testament, the fig is considered a sign of prosperity and blessing. The parable of the barren fig tree, told in the gospels, is an example of the use of the fig as a spiritual metaphor, emphasizing the importance of spiritual productivity in an individual's life.

The Symbolism of the Fig

The fig is often associated with notions of knowledge and wisdom. In the Judeo-Christian tradition,

The fig leaf is a symbol of original sin and man's awareness of his own vulnerability and imperfect nature. In Hindu tradition, the fig tree in pagodas is the place where Buddha achieved enlightenment, symbolizing the quest for spiritual truth.

The Fig as a Spiritual Metaphor

The growth of the fig from green to ripe fruit is often used as a metaphor for spiritual maturation and personal growth. Likewise, the fig evokes the duality of human experience, representing both sweetness and bitterness, joy and suffering, light and darkness. This symbolism reflects the complex nature of spiritual life and encourages understanding of the balance between opposing aspects.

The Fig and the Connection to Nature

In many spiritual traditions, the fig embodies the deep connection between man and nature. By honoring the fig, individuals recognize the wisdom of nature and the role each element plays in the balance of the universe. The fig thus becomes a reminder of the harmony between man, the earth and the sacred. The fig, as a religious and spiritual symbol, transcends cultural and temporal boundaries. It resonates with the human yearning for knowledge, wisdom, spiritual growth and connection to something greater than oneself. The symbolism of the fig reminds us that, just as the fruit ripens through the seasons, the human soul evolves and grows in its quest for meaning and understanding. The fig remains a tangible link between human experience and the divine, inviting everyone to meditate on the deep mysteries of life and spirituality.

Chapter 29: The Prominent Fig Trees: The Splendor of Fig Trees in Botanical Gardens

Renowned

Renowned botanical gardens around the world are havens of biodiversity and natural beauty, home to a staggering variety of plants from every corner of the planet. Among the green treasures that populate these gardens, the fig trees stand out for their majesty, their history and their symbolism. From one era to another, fig trees have thrived in botanical gardens, providing visitors with a captivating insight into the complexity and diversity of the plant world.

The Botanical Garden of Rio de Janeiro, Brazil

The Botanical Gardens of Rio de Janeiro, Brazil, are home to a Banyan fig tree (Ficus benghalensis) which spreads over a large area. This majestic tree creates an intricate network of aerial roots that wrap around the host tree, creating a visually stunning structure. Visitors can stroll under the beneficial shade of its tangled branches, discovering the magic of nature unfolding before their eyes.

The Singapore Botanic Gardens

The Singapore Botanic Gardens is home to a fascinating species of fig tree: the pagoda fig (Ficus religiosa). This variety is also known as "Bodhi Tree" and is revered for having been the tree under which Buddha achieved enlightenment. The tree at the Singapore Botanic Gardens is a descendant of the historic tree, thus establishing a spiritual connection with ancient wisdom.

Brooklyn Botanic Garden, United States

The Brooklyn Botanic Garden in New York has a Moreton Bay fig tree (Ficus macrophylla) that has an imposing presence. With its aerial roots draped in the air, this tree seems straight out of a fairy tale. Visitors are spellbound by how the tree has evolved to coexist with its environment, creating an extraordinary vision of adaptation and natural beauty.

The Majorelle Garden, Morocco

The Jardin Majorelle in Marrakech, Morocco, is famous for its exotic gardens, bright colors and enchanting atmosphere. Among the lush plants, the fig trees add a touch of mystery and authenticity. The garden is also home to a prickly pear cactus (Opuntia ficus-indica), a cactus with fleshy figs, symbolizing resilience and adaptation to arid environments.

The Importance of Fig Trees in Botanical Gardens

Fig trees are not just imposing trees in botanical gardens, they embody history, diversity and the deep relationship between man and nature. Their presence evokes ancient stories, cultural connections and symbolic meanings. Fig trees in renowned botanical gardens are more than just visual attractions; they are living ambassadors of the complexity and beauty of plant life, captivating visitors and inspiring a renewed respect for the natural world.

Chapter 30: The Advent of the Fig Tree: Cultivating the Future in the Face of Climate Change

Climate change, an undeniable reality, has profound impacts on agriculture and food production across the world. In this context, fig cultivation, which has a long history of relationship with humanity, faces new challenges and opportunities. By exploring sustainable and innovative approaches, fig cultivation could play a crucial role in adapting to climate change and preserving natural resources.

Climate Resilience

The fig tree is recognized for its ability to adapt to varied climatic conditions. However, climate change may alter patterns of temperature, precipitation and seasonality, which could impact the growth and fruiting of fig trees. Studies on heat and drought resistant fig varieties could be essential to ensure the sustainability of this crop.

Optimization of Water Resources

In a context of increasing water scarcity in many regions, water management becomes crucial for fig cultivation. Efficient irrigation methods, like drip irrigation, could minimize water waste while providing fig trees with the resources they need to thrive. The search for sustainable agricultural practices that reduce water demand while maximizing yields will be crucial for the future of the fig tree.

Genetic Diversity as an Ally

The genetic diversity of fig tree varieties is an invaluable resource for facing climatic challenges. By identifying and preserving varieties resistant to changing climate conditions, farmers can ensure the long-term sustainability of fig cultivation. Breeding programs aimed at developing climate-adapted varieties can strengthen the resilience of this crop.

The Transition Towards Sustainable Practices

The shift to sustainable agricultural practices is essential for the future of fig cultivation. Adopting agroecology, reducing the use of pesticides and chemical fertilizers, as well as promoting biodiversity in orchards, can help maintain balance ecological and to minimize negative impacts on the environment.

Education and Awareness

Educating farmers and local communities on climate change issues and best agricultural practices is crucial to ensure the sustainability of fig cultivation. Raising awareness of the importance of ecosystem preservation, reducing carbon emissions and managing natural resources can inspire positive actions.

The future of fig cultivation is closely linked to its ability to adapt to the challenges of climate change. By combining scientific innovation, sustainable agricultural practices and community outreach, it is possible to cultivate this ancient symbol in a new climatic context. Fig cultivation has the opportunity to become a model of adaptation and resilience in the face of changing climatic realities, while continuing to nourish future generations with its succulent fruits and rich symbolism.

Chapter 31: The Elegance of the Fig: Preservation of Biodiversity through the Ancient Tree

The fig, this sweet and succulent fruit, not only delights the taste buds; it also plays a crucial role in preserving biodiversity. As a keystone plant, the fig influences ecosystems by providing vital habitat for a variety of species, encouraging pollination and contributing to the fragile balance of nature. Beyond its delicious flavor, the fig is part of a complex network of biological interactions that nourish diversity and sustainability.

Ecological Habitat for Wildlife

Fig trees, with their bushy structure and abundant branches, create diverse habitats for many creatures, from small birds to insects and bats. Birds, in particular, find refuge in the branches and feed on the figs, thus contributing to the dispersal of seeds and the growth of new trees. Fig trees become sanctuaries where biodiversity flourishes.

Mutually Beneficial Relationships with Pollinators

Fig trees, often pollinated by specific wasps, establish unique symbiotic relationships. Male fig trees produce flowers that harbor pollinating wasps, while female fig trees produce the fruit. This subtle dance between trees and pollinators is an example of how biodiversity intertwines to ensure the reproduction and sustainability of species.

Promotion of Plant Biodiversity

Fig trees also play a role in promoting plant biodiversity. Their thick branches provide support for epiphytic plants, which grow above the ground without parasitizing the host tree. These plants add an extra layer of diversity in the ecosystem, creating an environment rich in species and habitats.

Significance for Indigenous Communities

In many regions, fig trees are revered by indigenous people for their roles ecological and cultural. Sacred fig trees are often considered focal points of biodiversity and are protected accordingly. They symbolize the harmonious relationship between man and nature, recalling the importance of preserving biological wealth for future generations.

The Responsibility of Conservation

The fig, as a key element of biodiversity, highlights the human responsibility to preserve ecosystems and protect species. Deforestation, climate change and other factors threaten these delicate interconnections. By taking steps to protect fig trees and the ecosystems to which they contribute, we honor the complexity of life and help maintain the stability of the planet.

The fig, with its fundamental role in biodiversity, is a reminder of how each natural element is woven into the complex web of life. From the smallest insect to majestic trees, each actor plays an essential role in preserving biodiversity. By appreciating the fig for more than its taste delicacy, we honor its contribution to the diversity of life and reinforce our commitment to protect and preserve the biological wealth that nourishes our planet.

Chapter 32: The Amazing Palette of Figs: A Dance of Colors and Shapes

Figs, iconic fruits with captivating flavors, not only charm our taste buds, they also amaze our eyes with their fascinating range of colors and shapes. From shimmering hues to varied silhouettes, figs provide a captivating visual spectacle that reflects the diversity of nature and awakens our appreciation for beauty in all its richness.

A Color Palette

Figs come in a stunning variety of colors that range from bright green to deep purple to golden yellow. Green figs are often the first to appear on the trees, announcing the start of the season. As figs ripen, they can take on darker, rich hues, ranging from dark purple, almost black, to burgundy red. Some varieties display even shades of brown and orange.

Elegant and Intriguing Shapes

Figs are not only attractive in their colors, they also have a diversity of shapes that add to their charm. Some figs are round and fleshy, while others are more elongated and tapered. Figs "drop of water" have a teardrop shape, while "Turkish" have a more rounded and flattened silhouette. This variety of shapes demonstrates the many facets of nature and adds to the visual tasting experience.

Natural Works of Art

Figs, with their variations in color and shape, are like natural works of art that evolve as they mature. Every nuance and contour tells a story of growth, transformation and cycles of life. By observing them, we are reminded of the magic of nature and the complexity of creation.

The Reflection of Natural Diversity

The diversity of colors and shapes of figs is a reflection of the natural diversity that characterizes our world. Each fig variety carries with it the history of its terroir, its environment and the forces that contributed to its unique growth. This diversity reminds us of the importance of preserving old and local varieties to maintain the genetic richness of plants.

Figs are more than just a treat for the senses, they are a visual celebration of nature's creativity. Their palette of colors and their varied shapes are all testimonies to the beauty that unfolds in the natural world. By savoring figs, we are invited to contemplate the visual symphony of nature and to renew our respect for the diversity that beautifies our planet.

Chapter 33: Fruity Fascination: Figs in Contemporary Popular Culture

Figs, those sweet and meaty delights, are not only a treat for the taste buds, but they have also found their way into contemporary popular culture. Whether through food, fashion, art or even social media, figs continue to capture people's imagination and spark lasting interest.

The Feast of Gastronomy

Figs have gained popularity in contemporary cuisine, found not only in traditional desserts, but also in a variety of savory and sweet dishes. From salads to fine cheeses, from toast to grilled meats, figs add a sophisticated touch to a multitude of recipes. Their combination of sweetness and bitterness offers a complex and delicious flavor palette, expanding the culinary horizons of food lovers.

Elegant Fashion

Figs have also taken over the fashion world. Their rich color palette, ranging from dark purple to deep red, has inspired designers to create clothing and accessories that reflect this alluring hue. From evening dresses to jewelry, figs have transformed into a source of visual inspiration for contemporary designers.

The Art of Creativity

Figs are not only eaten, they also inspire artists to create captivating visual works. From paintings to photographs, sculptures to illustrations, figs have become popular artistic subjects. Their unique shapes and colors provide a canvas for creative expression for contemporary artists seeking to capture the beauty of nature in their work.

Social Media: Virtual Showcase

Figs have also established their presence on social media, where food lovers, photography and lifestyle share their culinary and aesthetic creations. Hashtags dedicated to figs are flooding platforms, showing how this fruit has captured hearts and stomachs around the world. Food blogs, Instagram accounts and recipe videos have all helped elevate figs to icon status in contemporary popular culture.

Figs, an ancient fruit with rich symbolism, have managed to integrate harmoniously into contemporary popular culture. They embody a convergence between tradition and modernity, between taste pleasures and artistic expressions. From fine foods to artistic creations, figs continue to inspire, delight and beautify our lives, while celebrating their place in an ever-changing world.

Chapter 34: Natural Radiance: Fig in Modern Cosmetics

Modern cosmetics has opened its doors to a variety of natural ingredients with beneficial properties, and among them, the fig stands out for its potential to provide remarkable benefits to the skin and hair. From innovative formulations to beauty products, fig has become a rising star in the beauty world, offering a touch of natural elegance and refinement to skincare rituals.

Nutrition and Hydration

Figs are rich in essential nutrients, including vitamins, minerals and antioxidants. In cosmetics, these properties translate into intense nutrition and deep hydration for the skin and hair. Fig products help prevent dehydration, soothe dry skin and condition damaged hair, providing natural shine.

Antioxidants and Anti-Aging

The antioxidants found in figs play a crucial role in combating the effects of aging. They help protect the skin against free radicals and prevent premature signs of aging, such as fine lines and wrinkles. Fig-based beauty products provide natural support to the skin, promoting cell regeneration and maintaining a youthful, radiant appearance.

Gentle and Natural Exfoliation

Fig contains natural enzymes that help to gently exfoliate the skin, removing dead cells and revealing a brighter, even complexion. Fig-based exfoliating products offer a gentle, non-abrasive alternative to chemical treatments, while nourishing the skin and promoting cell renewal.

Hair Protection

For hair, figs provide natural protection against environmental damage, while strengthening hair structure and improving elasticity. Fig hair products help prevent split ends, add shine, and maintain overall hair health.

Commitment to Sustainability

The growing popularity of fig-based cosmetic products is also part of the movement towards more sustainable beauty. Natural and renewable ingredients like fig are seen as environmentally friendly and concerned with skin health. Consumers are looking for products that are both effective and planet-friendly, and fig ticks both of those boxes elegantly.

The fig, with its nutritional virtues and its regenerating properties, has won a place of choice in the world of modern cosmetics. From lotions to masks to serums, it brings a touch of luxurious naturalness to beauty rituals. While embracing tradition and science, the fig celebrates its role as an accomplice to skin and hair, providing both a pleasurable sensory experience and long-lasting benefits for beauty that radiates from the inside out. #39;exterior.

Chapter 35: Vigilance and Remedies: Managing Common Fig Tree Diseases

The fig tree, a symbol of fertility and wealth, is not without its plant health challenges. As

All plants, fig trees are subject to certain diseases which can compromise their growth and productivity. However, with a thorough understanding and proper prevention measures, it is possible to protect these precious trees from common ailments and keep them healthy.

Powdery mildew

Powdery mildew is a fungal disease that appears as a white powdery coating on the leaves, stems and fruits of the fig tree. To prevent and treat powdery mildew, it is important to maintain good air circulation around the tree by pruning densely leafy branches. Applying a sulfur fungicide can also help control this disease.

Fruit Rot

Fruit rot, often caused by fungi or bacteria, can affect figs, especially in humid weather. To prevent fruit rot, it is advisable to harvest ripe figs as soon as they are ready, handle them with care to avoid injury, and store them in a dry, well-ventilated area. Using appropriate fungicides can also help prevent this disease.

Rust

Rust is a fungal disease that appears as brown or red spots on the leaves of the fig tree. Rust prevention involves keeping leaves dry by avoiding watering the foliage. If infected, pruning the affected parts can help limit the spread of the disease. Copper fungicides can also be used to treat rust.

Bacterial Necrosis

Bacterial necrosis is a disease that causes dark brown or black lesions on the stems and branches of the fig tree. Preventing bacterial necrosis involves practicing careful pruning by removing infected parts, disinfecting tools between each cut to prevent the spread of bacteria. In cases of severe infection, it may be necessary to remove the infected tree to prevent disease to spread to other fig trees.

Managing common fig tree diseases requires continued vigilance and a proactive approach. The key lies in prevention, promoting a healthy environment and implementing appropriate growing practices. Early identification of symptoms and targeted use of treatment methods, such as fungicides and pruning practices, can play a vital role in maintaining the health of fig trees. By combining disease knowledge with proper precautionary measures, fig lovers can maintain the vitality of their trees and continue to enjoy these natural delights.

Chapter 36: Natural Balance: The Natural Enemies of the Fig Tree

In the complex ecosystem of the fig tree, a delicate harmony is maintained through the presence of natural enemies. As the fig tree grows and thrives, it is confronted with a variety of organisms which, although considered "enemies", play an essential role in maintaining the biological balance and health of the fig tree. #39;tree. These natural enemies are not only predators, but also regulators that contribute to the diversity and stability of the ecosystem.

Insectivorous Predators

Fig trees are home to a multitude of insects which, although they may seem pests, act as natural predators for other organisms which could cause damage. Spiders, ladybugs and parasitoid wasps are some of these insectivorous predators that feed on insect pests such as aphids and mites. By regulating insect pest populations, these predators help maintain the overall health of the fig tree.

The Birds and the Bats

Fig trees produce abundant fruit which, in addition to feeding humans, is a source of food for many animals. Birds and bats feast on ripe figs, helping to disperse the seeds and encourage the growth of new trees. In exchange, these animals help also to control populations of harmful insects by feeding on species that could damage the fig tree.

Plant Biodiversity

The presence of a diversity of plants around the fig tree can also contribute to the tree's natural protection. Some plants produce chemical compounds that repel harmful insects or attract natural predators. By promoting plant biodiversity, fig tree owners can create an environment favorable to biological regulation and the prevention of infestations.

The Precious Balance

The coexistence of fig trees with their natural enemies reflects the complex and subtle balance that characterizes natural ecosystems. These natural enemies, often considered pests at first glance, are actually the guardians of balance, ensuring that the growth of the fig tree does not become uncontrolled and that no one organism becomes too dominant. The absence of these natural enemies could disrupt the food chain and lead to unwanted imbalances.

The presence of natural enemies in the fig tree ecosystem is a poignant reminder of the complexity of life and the interdependence of species. The interactions between fig trees, predators and surrounding plants form a web of interconnections that promote ecosystem diversity, health and sustainability. By respecting this natural balance, we celebrate the richness of nature and promote the harmonious coexistence of all creatures that share the world of fig trees.

Chapter 37: Naturally Delicious: Figs in Vegan Cooking

Vegan cuisine, characterized by its respect for living beings and the environment, thrives on a diverse range of plant-based ingredients. Among them, figs shine as a source of natural deliciousness, bringing a sweet and nutritious touch to vegan dishes. From starters to desserts, figs offer gastronomic versatility that fits perfectly with the ethics and flavors of the cuisine vegan.

Delicacy in Savory Dishes

Fresh or dried figs bring a subtle, contrasting sweet note to savory dishes, creating a balance of flavors that delights the taste buds. They can be used in salads to add a touch of sweetness, in whole grain dishes to create a rich taste experience, or even in sauces to create a sweet and tangy base.

Figs in Vegan Desserts

When it comes to desserts, figs are undisputed stars. They can be made into compotes, jams or fillings for vegan cakes. Dried figs, when rehydrated, become a naturally sweet treat to add to muffins, cookies and other sweet treats.

Vegetable Cheeses and Figs

A classic combination in vegan cuisine is vegetable cheeses and figs. Figs pair perfectly with hard or soft plant-based cheeses, adding a sweet, textural touch that mimics the experience of traditional cheeses. These pairings create an explosion of flavors that delight vegan palates.

Energy and Natural Nutrition

Figs, rich in fiber, vitamins and minerals, offer a nutritional boost to vegan cooking. Their nutritional properties make them an ideal choice for vegan recipes that aim to provide a sustainable source of energy while meeting essential nutrient needs.

Ethics and Culinary Creativity

The use of figs in vegan cooking reflects the commitment to ethical eating and environmentally friendly. Figs, being natural, non-animal products, fit perfectly into the principles of vegan cooking. Plus, they inspire culinary creativity, giving vegan chefs a canvas of flavors on which they can paint culinary masterpieces.

Figs, symbols of fertility and sweetness, fit harmoniously into vegan cuisine, providing a natural touch of delicacy and sweet flavor. Their versatility makes them valuable ingredients for savory and sweet dishes, starters and desserts. Figs are not only allies for the taste buds, but also for the ethical and environmental values of vegan cuisine. By incorporating them with creativity and passion, fans of vegan cuisine can treat their palates to a culinary experience that celebrates nature, health and respect for all forms of life.

Chapter 38: The Art of Grafting Fig Trees: Merging Nature with Technology

Grafting, an ancestral technique for propagating plants, has developed into a refined art over the centuries. When applied to fig trees, this technique takes on a new dimension, allowing enthusiasts to create unique varieties, restore ancient trees and share their love for these majestic trees. The art of grafting on fig trees is a demonstration of the harmonious collaboration between the hand of man and the power of nature.

The Fusion of Two Individuals

Grafting involves the fusion of a rootstock, which provides roots and support, with a scion, which provides the desired characteristics of the variety. In the case of fig trees, this fusion creates a new harmony between the vigor of the rootstock and the distinctive features of the scion. Grafts make it possible to quickly multiply exceptional varieties and preserve rare or old specimens.

Grafting Techniques

Several grafting techniques are used on fig trees, each suited to specific purposes. Slit grafting, escutcheon grafting and inlay grafting are among the commonly used methods.

employees. Each of these techniques requires careful precision and a deep understanding of fig tree physiology.

The Creation of New Varieties

The art of grafting fig trees allows horticulturists and hobbyists to create new varieties by combining desired characteristics of different fig trees. For example, a fig tree that produces exceptionally sweet fruit can be grafted onto disease-resistant rootstock. This creative approach opens the door to exploring unique flavors and visual aspects.

Preservation of Plant Heritage

Old and rare fig trees may be threatened by factors such as disease, environmental changes or neglect. Grafting then becomes an essential tool for preserving these precious specimens. Grafting a piece of an ancient fig tree onto healthy rootstock ensures the survival of unique traits and ancient stories.

Patience Rewarded

Grafting fig trees requires a generous dose of patience. Results aren't instantaneous, but the time invested results in lasting rewards. Properly done grafts can produce vigorous, productive fig trees, creating a living legacy for future generations.

The art of grafting on fig trees embodies the fusion of science, technique and creativity. It serves as a reminder that human hands can work in harmony with the forces of nature to create something new while respecting the roots of the past. By mastering this technique, fig tree enthusiasts enrich the history of these exceptional trees and contribute to the preservation and diversity of these plant gems. Grafting fig trees is much more than a technique, it is a celebration of life, growth and the art that connects man to the earth.

Chapter 39: Sweet Symbols: Traditional Festivities Celebrating Figs

Figs, succulent fruits steeped in symbolism and history, are celebrated across the world in traditional festivities. These joyful events bring fig lovers together to honor this precious fruit, not only for its exquisite taste, but also for the cultural significance it carries. From ancient rituals to modern feasts, festivities celebrating figs are a living tribute to the richness and diversity of human culture.

The Blessed Harvests

In many cultures, figs are harvested at specific times of the year, and these harvest periods are often marked by religious or agricultural festivities. Figs are harvested with care, and in some places the first harvest is honored with special prayers and ceremonies. These festivities demonstrate the deep relationship between man and nature, and the importance of figs in sustenance and culture.

Fig Festivals around the World

In Turkey, the Golden Fig Festival is a celebration that highlights agricultural and gastronomic traditions linked to figs. In Morocco, the Fig Festival in Bouznika is an opportunity for farmers and fig lovers to come together and exchange experiences. In Italy, the village of Solopaca celebrates the Fig Festival, where fig products and local specialties are featured. These festivities reflect how figs are integrated into the culture of different regions.

Gastronomy and Creativity

Festivities celebrating figs highlight the gastronomic richness of this fruit. Local chefs and culinary enthusiasts compete in ingenuity to create a variety of dishes and desserts highlighting figs. From jams to pastries, savory dishes to drinks, figs are in the spotlight in all their forms, capturing the imagination of taste buds and inspiring unique culinary creations.

Art, Music and Dance

Some traditional festivities celebrating figs go beyond gastronomy to encompass artistic expressions. From art exhibitions featuring works inspired by figs to musical performances and folk dances, these festivities provide a rich and vibrant cultural backdrop. Figs thus become a source of inspiration for artists and creators.

Cultural Transmission

These traditional festivities are not only limited to celebrating figs, but also play a role in cultural transmission and heritage preservation. They allow future generations to connect with ancient practices and values, thus strengthening the link between past and present.

Traditional festivities celebrating figs are an ode to culture, nature and the rich heritage that these fruits carry within them. They embody the deep connection between man and the earth, between the terroir and the table. By celebrating figs through these events, communities honor a sweet symbol that transcends taste to become an integral part of their cultural identity.

Chapter 40: The Sweet Muse: The Fig as a Source of Artistic Inspiration

Since ancient times, the fig has captured the imagination of many artists, poets, painters and writers. This fleshy, delicate fruit has transcended its sweet nature to become a muse in the art world. Its elegant form, shimmering colors and rich symbolism have inspired artistic creations that celebrate beauty, mystery and sensuality.

Evocative Paintings

The fig has often featured in paintings throughout the ages, from ancient art to modern works. Still lifes, in particular, allowed artists to explore the complex shape and textures of figs. The vibrant details of their skin, the fleshy softness of their interior, have been reproduced with a meticulousness that testifies to the admiration for this fruit.

Symbolism on Canvas

The fig, with its connotations of fertility, sensuality and pleasure, has become a powerful symbol in art. It has been used to represent themes such as abundance, temptation and the fleeting nature of life. In religious and mythological works, figs have sometimes been used to infuse deeper meaning into stories.

Literary Inspiration

The fig has also found its way into literature, where it has been sung about by poets and writers for its beauty and symbolism. It was used as a metaphor to express the sweetness of life, seduction or even transformation. The fig, with its luxurious texture and intoxicating flavor, nourished not only the body, but also the imagination of the authors.

Culinocentric Creativity

Figs have not only inspired visual and literary works, they have also been the source of inspiration for culinary creators. The artist chefs have designed visually stunning dishes that showcase the color palette and shape of the figs. Artistic cuisine has transformed figs into edible masterpieces, combining taste with aesthetics.

The Fig in Contemporary Art

Today, the fig continues to inspire contemporary artists. From sculptures to photographs to art installations, figs are explored from new and inventive angles. Modern art often expresses a complex relationship with nature and food, and the fig offers a rich starting point for these explorations.

The fig transcends its status as a delicious fruit to become a rich and timeless source of artistic inspiration. Its sensual form, evocative colors and deep symbolism have inspired visual, literary and culinary works of art throughout the ages. As a sweet muse, the fig continues to invite artists to explore the multiple dimensions of beauty, symbolism and creativity in the world of art.

Chapter 41: A Balanced Ecosystem: Fig Trees in the World of Permaculture

Permaculture, a holistic approach to ecological design, aims to create sustainable and balanced systems by drawing inspiration from natural patterns. Fig trees, with their ability to thrive in a variety of conditions, play a vital role in permaculture designs. By integrating fig trees into these systems, permaculture practitioners benefit from their contribution to biodiversity, soil regeneration and ecosystem resilience.

Fig Trees as Pivot Plants

In permaculture systems, fig trees can be used as pivot plants. Their large leaves provide shade and create a favorable microclimate for other plants growing at their base. Depending on design needs, fig trees can be strategically positioned to provide shade for heat-sensitive crops or to create thermal regulation zones.

Reducing Soil Erosion

The deep, sturdy roots of fig trees act as anchors, helping to stabilize soils and reduce erosion. Fig trees can be integrated into permaculture designs to protect vulnerable soils from leaching caused by rainfall. By strengthening soil integrity, fig trees promote overall ecosystem health.

Benefits for Biodiversity

Permaculture promotes biodiversity by creating balanced ecosystems. Fig trees, by attracting a variety of insects, birds and small mammals, contribute to biological diversity. Figs also serve as a food source for these creatures, strengthening the connections between different elements of the ecosystem.

Natural Fertilization

Fig trees are known for their ability to grow in relatively poor soils. By spreading their deep roots to reach nutrients, they extract mineral elements which are then redistributed when the leaves fall and decompose. This natural fertilization improves soil fertility and benefits neighboring plants.

Sustainability in Designs

When creating permaculture designs, fig trees can be used to maximize mutual benefits between elements of the system. For example, they can be strategically placed to provide shade to water collection areas, helping to reduce evaporation and support water retention in the soil.

Fig trees embody the fundamental principles of permaculture as elements that enhance diversity, soil regeneration and ecosystem sustainability. Their ability to provide shade, stabilize soil and promote biodiversity makes them valuable in the design of permaculture systems. Incorporating fig trees into these designs is to embrace the philosophy of permaculture by creating balanced systems that imitate and interact harmoniously with nature.

Chapter 42: Tangled Legends: The Mysterious Urban Legends Around the Fig Tree

Fig trees, majestic trees full of symbolism and sweet flavors, have fueled the human imagination for centuries. Into the fabric of cities and urban spaces, fig trees have also woven spellbinding legends. Between nocturnal rumors and stories passed down from generation to generation, these urban legends transport the mystery of the fig tree in stories that are woven into the fabric of urban life.

The Haunted Fig Tree

Some urban legends surround the fig tree with frightening mysteries. It is said that certain fig trees, particularly old and isolated, are haunted by spirits or ghosts. The twisted branches and

Shadows cast by the glow of the moon can fuel stories of supernatural encounters under the fig trees. These stories, shared by candlelight during all-night vigils, capture the mysterious ambiance of the urban night.

Vows and Secrets

Fig trees, with their solemn and imposing nature, have inspired legends about their ability to hear and keep secrets. It is said that if one whispers a wish or wish to a fig tree, it will come true. These legends add a touch of magic to urban fig trees, inviting passersby to entrust their deepest hopes and desires to these caring trees.

Wrecks of the Past

Some urban fig trees have been standing for decades, even centuries. Their deep roots lived in times gone by, and legends emerged around these "silent witnesses" of urban history. Fig trees are said to hide long-buried secrets, from lost treasures to forgotten stories, making them guardians of the urban past.

The Mysterious Creature

The shadows cast by the branches of fig trees at night have inspired stories of mysterious creatures hiding among the leaves. Urban legends describe strange beings, half-human, half-folk, who emerge from the fig trees to wander the dark alleys. These stories, although improbable, contribute to the sense of wonder and strangeness in the urban environment.

Bewitching Beauty

Fig trees, with their dense leaves and impressive shapes, are often described as having a haunting beauty. Urban legends suggest that those who gaze upon the majesty of a fig tree under a full moon can be bewitched by its power. These tales reflect how fig trees, with their imposing presence, can capture the imagination and catch the eye of passers-by.

Urban legends surrounding fig trees reveal the power of the human imagination and the ability of these trees to blend into the urban fabric, while retaining an aura of mystery. These stories, passed down through generations, enrich the relationship between city dwellers and fig trees, making trees creators of stories as well as elements of the landscape. In the labyrinth of urban legends, fig trees remain keepers of secrets and catalysts of wonder.

Chapter 43: Mediterranean Figs: Tasty Relics of Cultural Heritage

Figs, with their enchanting sweetness and luxurious texture, have been intimately linked to Mediterranean culture for millennia. In this sun-drenched region, fig trees flourished and shaped the cultural landscape in profound ways. From sacred symbols to sumptuous feasts, figs in Mediterranean culture embody a rich history, tradition and abundance that transcends borders.

Fertile Ancestors

Figs are often associated with fertility, and in Mediterranean culture they embody the abundance of the bountiful earth. In many ancient civilizations, figs were considered a gift from nature, a sign of blessing from Mother Earth. This connection between figs and fertility has endured, shaping festivities and celebrations.

Figs and Spirituality

In many Mediterranean cultures, fig trees are linked to spiritual and religious practices. Fig trees are mentioned in religious texts and are often associated with wisdom, patience and perseverance. They have been seen as symbols of spiritual transformation and connection with divine forces.

Feasts and Culinary Traditions

Figs occupy a place of honor in Mediterranean cuisine. Fresh or dried, they are used in a variety of dishes, from appetizers to desserts. Cheese-stuffed figs, fig tarts and fig jams are popular culinary delights in the region. Mediterranean feasts are often adorned with dishes showcasing the richness and flavor of figs, adding a touch of rustic sophistication.

Crafts and Customs

Mediterranean fig trees have also found their place in local crafts and customs. Fig leaves, for example, have been used to wrap and cook traditional dishes, like dolmas. The ubiquity of fig trees in the landscape has also influenced local art and architecture, adding a cultural dimension to the relationship between people and trees.

Figs and Social Events

Figs played an important role in Mediterranean social events and community gatherings. Fresh or dried figs are often offered to guests as a sign of warm hospitality. Weddings, religious festivals and family celebrations are enhanced with fig-based dishes, creating social bonds through shared tasting.

Figs in Mediterranean culture are much more than just food. They embody the history, beliefs, traditions and abundance of a region rich in cultural diversity. Fig trees, with their beneficial shades and sweet fruits, are silent witnesses to a harmonious relationship between man and nature. Revealing the richness of Mediterranean culture, figs continue to serve as a bridge between past and present, between land and table, while evoking the sweet taste of cultural heritage.

Chapter 44: Literary Delights: The Fig in Contemporary Literature

In contemporary literature, the fig has transformed into a flavorful and multidimensional metaphor, symbolizing both sensory pleasure and emotional depths. Authors

Contemporary artists explore the fig from multiple angles, associating it with themes such as sensuality, nostalgia, the quest for self and the links between man and nature. Figs in contemporary literature are much more than just a fruit: they embody layers of emotions and meanings that add a rich and complex dimension to modern stories.

Eroticism and Sensuality

The fig has long been linked to sensual associations, and in contemporary literature it continues to play this role. The authors explore the velvety and fleshy textures of figs to evoke erotic sensations and intense sensory experiences. The descriptions of ripe, juicy figs become subtle metaphors for moments of passion and desire.

Memory and Nostalgia

Figs, with their rich flavor and intoxicating sweetness, are often used to evoke memories and moments from the past. Contemporary authors use figs to create nostalgic vignettes, transporting readers to scenes from childhood, youth or bygone eras. Figs become portals to emotionally charged memories and reflections on the passing of time.

The Quest for Self and Identity

In some contemporary stories, the fig is used as a metaphor for the quest for self and the discovery of identity. The fig, with its hidden interior and protective skin, reflects human complexity and deep layers of the soul. Literary characters often find themselves through a fig-like exploration of the self, unveiling hidden facets over time.

Relationship with Nature

The fig in contemporary literature is sometimes used to explore the relationship between man and nature. Contemporary authors examine how figs, rooted in terroir and shaped by natural elements, represent a deep connection to the natural world. This exploration

Symbolism reveals how nature can influence our understanding of ourselves and our emotions.

Figs in contemporary literature transcend their status as simple fruits to become complex symbols that feed the readers' imagination. By associating figs with themes as varied as eroticism, nostalgia, the quest for self and the relationship with nature, contemporary authors breathe depth and emotional richness into modern stories. Figs become bearers of meaning, tools for exploring human complexity and for weaving links between individual and universal experiences.

Chapter 45: Regional Delights: The Fig in Regional Culinary Practices

Figs, rich in sweetness and flavor, have long found their place at the heart of regional culinary practices around the world. In different regions, figs have been ingeniously incorporated into traditional dishes, from exquisite confections to elaborate savory dishes. Each culture has added its unique twist to the way it appreciates and celebrates this delicious fruit. Figs in regional culinary practices embody the subtle fusion of nature and culture, creating delights that tell stories of terroir and tradition.

Mediterranean: A Feast of Colors and Flavors

The Mediterranean region has a deep relationship with figs, which is reflected in its cuisine. From fresh figs to dried figs, they are found in a variety of dishes. In Greece, figs are often used in desserts, such as baklava or honey-dried figs. In Italy, fresh figs are sometimes served with cheese, creating an exquisite blend of sweet and savory. Figs are also used to enhance meat or fish dishes, adding a sweet-umami dimension.

Middle East: Oriental Sweets

Figs also play a central role in Middle Eastern culinary traditions. Stuffed figs with nuts and honey, like ma'amoul, are sweets popular during celebrations and special events. Dried figs are also used to add a touch of sweetness to meat dishes, creating a harmony of flavors. Figs in Middle Eastern cuisine embody the refinement and complexity of the region's flavors.

Asia: Yin and Yang Balance

In Asia, figs are often seen as having beneficial health properties. Figs are used in traditional Chinese medicine and are also incorporated into cooking. In Korea, figs are sometimes pickled to accompany main dishes or are used in refreshing drinks. Figs in Asia embody the balance between nutrition and delight.

Latin America: Fusion of Flavors

In some parts of Latin America, figs are used to add an exotic touch to traditional cuisine. Figs can be used to garnish salads, add sweetness to spicy dishes, or made into jams and preserves. The fusion of flavors resulting from combining figs with local ingredients creates unique taste experiences.

Figs, rich in history and flavor, play an essential role in regional culinary practices around the world. Every culture has brought creativity to the use of figs, creating a mosaic of sweet and savory delights. Figs embody the fusion between nature and human creativity, offering a diversity of tastes that reflect the richness of terroirs and traditions. From the Mediterranean to Asia, from confectionery to main dishes, figs in regional culinary practices are an invitation to explore the world through the prism of flavors.

Chapter 46: Spiritual Awakening under the Fig Trees: Eastern Spirituality and the Fig Trees

Fig trees, with their peaceful shadows and nourishing fruits, have forged deep connections with Eastern spirituality since time immemorial. In Eastern spiritual traditions, fig trees are become symbols of meditation, wisdom and awakening. Under their calming shade, profound teachings were imparted, meditations were practiced, and souls found a connection with the divine. Fig trees, revered as witnesses of enlightenment, embody the quest for truth and inner search.

The Tree of Enlightenment: Buddhism

One of the most iconic examples of the relationship between fig trees and Eastern spirituality is found in Buddhism. It was under a fig tree, the famous Bodhi tree, that Siddhartha Gautama achieved enlightenment to become the Buddha. Under the branches of this sacred fig tree, Siddhartha meditated deeply, transcending human suffering to find peace and truth. Today, Bodhi fig trees are revered in Buddhism as places of contemplation and spiritual awakening.

Wisdom in the Shadow: Hinduism

In Hinduism, fig trees are also associated with spirituality and wisdom. Fig trees are often mentioned in ancient texts like the Vedas and Upanishads, symbolizing the connection between heaven and earth. Some Hindu myths say that deities chose fig trees as their residences, providing the trees with an aura of sanctuary and knowledge.

The Protection of the Teachings: Japan and Zen

In Japan, Zen spirituality also has a deep relationship with fig trees. Ginkaku-ji Temple, or Silver Pavilion, is surrounded by fig trees that embody the simplicity and depth of Zen. Fig trees, with their delicate leaves and twisted trunks, are considered guardians of Zen teachings, reminding practitioners of the beauty of the present moment.

The Fig as a Spiritual Portal

In many Eastern traditions, fig trees act as portals to the divine and spiritual awakening. Their abundant nature, their generous shade and their nourishing fruits transform them into places conducive to meditation and contemplation. Fig trees are revered not only for their natural beauty, but also for their ability to provide a space for connection with the divine.

Fig trees and Eastern spirituality are intertwined in a deep and meaningful relationship. Under their branches, believers found enlightenment, inner peace and connection with the divine. Fig trees embody the quest for truth and wisdom, inviting souls to meditate, reflect and find enlightenment.

Rooted in centuries-old traditions, fig trees are much more than trees: they are living symbols of the human aspiration to transcend material limitations and embrace the deep spirituality of the soul.

Chapter 47: Hidden Treasures: The Rarest Varieties of Figs

Among the countless varieties of figs that populate the world, some stand out for their rarity and uniqueness. These rare fig varieties are botanical treasures, jewels of natural diversity. Each with its own characteristics and flavors, these rare figs remind us of the richness and variety of nature. In this chapter, we will explore some of the rarest fig varieties, which spark wonder among fruit lovers and gastronomic connoisseurs.

1. Greek fig tree

The Greek fig tree, also known as 'Vasilika', is one of the rarest and most valuable varieties of figs. Originally from the Mediterranean region, this fig is characterized by its bright green color and oblong shape. Its sweet flavor, combined with slightly lemony notes, makes it a rare and sought-after delicacy.

2. Djebba fig tree

The Djebba fig tree is a rare variety native to Tunisia. Its figs are distinguished by their deep purple hue and their dense, sweet flesh. The Djebba fig is appreciated for its delicate and sweet flavor, which evokes the Tunisian terroir.

3. San Pedro

The fig variety "San Pedro" is a rarity native to California. These figs are distinguished by their conical shape and dark purple color. They are known for their sweetness and juicy texture, creating a taste experience that evokes balmy California summers.

4. Bethlehem

The fig "Bethlehem" is a rare variety native to Palestine. These figs are small, with dark purple skin and pinkish flesh. The "Bethlehem" is prized for its sweet flavor and exquisite aroma, which evokes the peaceful hills of the Holy Land.

5. Variegated

The variegated fig, also called 'Tiger', is a rare and visually impressive variety. Its skin is striped with green and yellow hues, creating a unique tiger-like pattern. This variety is appreciated for its sweetness and velvety texture.

6. Dottato

Originally from Italy, the "Dottato" is rare and exceptional. Its figs are medium sized and have light green skin speckled with white dots. The "Dottato" is appreciated for its deliciously sweet flesh and floral aroma.

The rarest varieties of figs are botanical treasures that delight the senses and remind us of the infinite diversity of nature. Their unique flavors and distinctive visual characteristics make them unique creations of nature. Each rare fig variety evokes the history of its terroir and embodies the passion of growers for growing extraordinary fruits. These rare figs, while being delights for the taste buds, are also witnesses to the rich biodiversity and beauty that surround us.

Chapter 48: Sweet Treats: The Fig and the Confectionery Industry

In the enchanted world of confectionery, the fig has found its place among the sweet delights that make the eyes of gourmands shine. Its combination of lush flavor and delectable texture makes it a valuable component of many confectioneries around the world. From the sweetness of dried figs to the richness of filled figs, the fig has conquered the confectionery industry as a flagship ingredient. The fig has managed to charm the taste buds and bring its unique touch to the sweet world of confectionery.

A Coating of Softness

Dried figs, naturally rich in sweet flavor, are a staple in the confectionery industry. They are often coated in chocolate, caramel or sugar to create irresistible bites. Chocolate-covered figs, for example, strike the perfect balance between the sweetness of the fruit and the bitterness of the chocolate, creating an explosion of flavors in every bite.

The Art of Fur

Filled figs are a delicious expression of the confectionery industry's creativity. By combining figs with various toppings, such as nuts, dried fruits, spices or liqueurs, confectioners create sumptuous creations. Filled figs are often presented as little treasures, wrapping a tasty surprise inside.

Tradition and Innovation

In certain regions, the fig is at the heart of ancient sweet traditions. Stuffed figs, which may contain nuts, citrus or spices, are traditional sweets in many cultures. However, the confectionery industry is constantly innovating by introducing modern and creative twists. From matcha-coated figs to sea salt caramel-filled figs, confectioners are pushing the boundaries of creativity while celebrating the fruit's rich history.

A Global Gastronomic Journey

The fig, with its versatility and distinctive flavor, crosses gastronomic boundaries and finds its place in kitchens around the world. From sweet Middle Eastern delights, like fig baklava, to refined French confections where figs are incorporated into intricate pastries, the fig offers an endless variety of creative possibilities for confectionery artisans.

The fig and the confectionery industry combine harmoniously to create treats that evoke a unique sensory experience. Coated dried figs, filled figs and innovative creations transport the taste buds on a sweet journey through the world of confectionery. In this delectable union, the fig reveals its capacity to amaze and inspire, adding a touch of refined sweetness to the sweet world of indulgence.

Chapter 49: Fig Leaves: Natural Healers of the Past

For generations, fig leaves have been used as a valuable ingredient in traditional remedies across the world. Loaded with beneficial compounds, these leaves have become allies in the quest for natural healing. Their use in diverse cultures reveals a legacy of medicinal wisdom that endures to the present day. Fig leaves have found their way into traditional remedies, providing an array of health benefits and a deep connection with nature.

Ancient Healing Practices

The use of fig leaves for medicinal purposes dates back to ancient times. Ancient civilizations, such as the Egyptians, Greeks, and Romans, knew of the healing properties of these leaves and used them to treat various ailments. Fig leaves were often made into ointments, infusions or poultices to relieve pain and ailments.

Antioxidant Powers

Fig leaves are rich in antioxidants, which play a crucial role in protecting cells from oxidative damage. These antioxidants help strengthen the immune system, prevent premature aging and reduce the risk of chronic diseases.

Diabetes Management

Fig leaves have also been linked to diabetes management. Studies have shown that compounds found in fig leaves may help regulate blood sugar levels by improving insulin sensitivity. Fig leaf extracts are sometimes used as a natural supplement in the management of this condition.

Anti-Inflammatory Properties

The anti-inflammatory properties of fig leaves make them a popular choice for relieving various inflammatory conditions, such as joint pain and skin inflammation. Poultices made from fig leaves can help soothe irritation and promote healing.

Digestion and Intestinal Health

In some traditions, fig leaves have been used to support digestion and promote gut health. The anti-inflammatory and soothing properties of the leaves may help soothe an upset stomach and support healthy digestion.

Natural Revelation

The use of fig leaves in traditional remedies illustrates the healing power and wisdom of nature. The knowledge passed down from generation to generation has made it possible to discover the treasures of benefits hidden in these leaves. From managing chronic conditions to promoting overall well-being, fig leaves offer a striking example of the symbiosis between man and nature.

Fig leaves, with their exceptional medicinal potential, are a reminder of the richness of traditional remedies and ancient wisdom. Their use in various remedies reveals a deep understanding of the beneficial properties of the plants around us. Fig leaves have continued to evolve in the contemporary medicinal landscape, testifying to the fact that nature, with its precious gifts, remains an invaluable source of healing and well-being for humanity.

Chapter 50: Fusion of Flavors: Fig in Modern Fusion Cuisine

Modern fusion cuisine is a culinary symphony that blends diverse traditions, ingredients and culinary techniques to create new and daring taste experiences. At the heart of this culinary creativity is the fig, a fruit that has crossed gastronomic boundaries to blend harmoniously into fusion dishes. By combining the rich flavors and lush textures of the fig with diverse culinary influences, modern fusion cuisine is enriched with new taste dimensions. The fig is becoming the shining star in the culinary world of fusion.

A Palette of Possibilities

The fig offers an endless palette of possibilities for fusion chefs. Its natural sweetness pairs perfectly with sweet ingredients like caramel and honey, while contrasting beautifully with more robust flavors like aged cheeses and spicy meats. It is this subtle play of contrast that allows the fig to blend into fusion dishes, creating surprising harmonies for the taste buds.

Balance of Flavors

Modern fusion cuisine often relies on the delicate balance of flavors. Fig, with its sweet flavor and slightly earthy character, brings a unique element to this equation. It can be used to sweeten spicy dishes or to add a touch of refinement to savory dishes. Figs in modern fusion cuisine serve as a balancing agent, adding a sweet, complex nuance to culinary compositions.

Innovative Culinary Creations

Modern fusion chefs are constantly pushing the boundaries of creativity, and the fig is often at the heart of their bold creations. From fig seafood tartars to fig sushi, and more by caramelized figs on exotic pizzas, the fig becomes a blank canvas for chefs eager to experiment with new combinations.

Cultural and Gastronomic Connection

Modern fusion cuisine transcends borders, celebrating the world's cultural and culinary diversity. The use of fig in fusion cuisine speaks to its ability to connect to a variety of traditions and tastes. It adds a Mediterranean touch to an Asian dish or an exotic note to a European creation, strengthening the links between culinary cultures and creating a multisensory experience.

The fig, with its exceptional flavor and versatility, has become a key piece in the creative puzzle of modern fusion cuisine. By adding a sweet and complex dimension to fusion dishes, the fig plays the role of a true culinary acrobat, capable of bringing an exquisite touch to a variety of creations. In the ever-changing culinary landscape, the fig remains an endless source of inspiration for daring chefs and adventurous foodies.

Chapter 51: Religious Festivals and the Deep Symbolism of the Fig Tree

The fig tree, with its lush branches and delicious fruits, holds a special place in the symbolic landscape of religious holidays across different cultures and beliefs. Its presence in these celebrations goes beyond the simple botanical aspect to take on a deep spiritual meaning, evoking themes such as spiritual growth, connection with the divine and inner transformation. The fig tree has become a powerful symbol in religious festivals and embodies essential spiritual values.

Symbol of Spiritual Growth

The fig tree, with its slow but steady growth, has become a symbol of spiritual growth in many religious traditions. It is often associated with the patience and perseverance necessary to develop a deep relationship with the divine. The biblical story of Jesus cursing the barren fig tree, by example, highlights the importance of spiritual productivity and ever-evolving faith.

The Metaphor of Transformation

In some beliefs, the fig tree is considered a metaphor for inner transformation. The stages of fig tree growth, from flowering to fruit ripening, are interpreted as a mirror of the phases of spiritual life. As the fig tree evolves from seed to harvest, believers are encouraged to evolve from ignorance to spiritual knowledge.

Reconciliation and Spiritual Fertility

The fig tree is also linked to themes of reconciliation and spiritual fertility. In some traditions it is considered a symbol of reconciliation between man and God, evoking the idea of renewal and forgiveness. Figs, with their sweetness and abundance, are sometimes interpreted as symbols of spiritual fertility, representing the fruitful production of virtues and good deeds.

The Connection with Divinity

The fig tree has often been associated with connection with divinity. In Hinduism, for example, the "Peepal" is revered as a sacred tree, closely linked to the god Vishnu. Fig trees are sometimes planted near places of worship to symbolize the divine presence and communication between heaven and earth.

The fig tree, with its natural beauty and deep symbolism, has found its place at the heart of religious festivals à worldwide. Its role transcends the botanical to become a living metaphor for spiritual growth, inner transformation and connection with the divine. By evoking patience, reconciliation and spiritual fertility, the fig tree adds a deep and meaningful dimension to religious celebrations, reminding believers of the essential values of their faith and encouraging them to continue their spiritual journey.

Chapter 52: Natural Writing: Fig Trees in the Art of Calligraphy

The art of calligraphy, a profound and elegant artistic expression, transcends linguistic boundaries to capture the beauty of words through form and movement. In this universe of artistic lines and curves, fig trees find their place, bringing a deep connection with nature and rich symbolism. The fig tree, with its organic shapes and spiritual meaning, has inspired calligraphers to incorporate its essence into their works. Fig trees enrich the art of calligraphy by adding a natural and spiritual touch to this ancient art.

The Balance between Form and Meaning

Calligraphy is the art of transforming words into visually captivating works. Fig trees, with their distinctive silhouette, add an organic dimension to this art. The graceful curves of the leaves and the complexity of the structure of the fig trees are all aesthetic features that calligraphers integrate into their compositions. This symbiosis between natural forms and written words creates a harmonious balance between meaning and aesthetics.

The Symbolism of the Fig Tree

Fig trees have deep spiritual meaning in many cultures, making them all the more valuable in calligraphy. The fig tree is often associated with spiritual growth, patience and inner transformation. Calligraphers incorporate these meanings into their works, creating compositions that transcend simple writing to evoke universal themes.

The Incorporation of Nature

Nature, with its beauty and diversity, has always been a source of inspiration for artists. Fig trees, with their connection to earth and sky, add a touch of nature to calligraphy art. The details of the leaves, the contours of the branches and the symbolism of the fig trees provide an organic dimension that echoes both human creativity and natural creation.

A Bridge between the Spiritual and the Visual

Fig trees in calligraphy act as a bridge between the spiritual and the visual. They embody intangible values and express them through tangible forms. Calligraphers use fig trees to channel the spiritual meaning of words into visually powerful compositions, creating works that touch the soul and the senses.

Fig trees in calligraphy art are an eloquent example of how nature and spirituality merge to create a rich and profound aesthetic. By adding organic and symbolic elements to the art of calligraphy, fig trees enrich this age-old form of artistic expression. They remind observers of the beauty of nature, inner growth and spiritual meaning, making each calligraphic composition a work of art that transcends words to tell a visual and spiritual story.

Chapter 53: Sweet and Sweet Tales: The Fig in Tales and Legends for Children

Tales and legends for children, full of adventure, learning and magic, often draw on the richness of the natural world to weave their stories. Among the botanical treasures that embellish these stories, the fig occupies a special place. With its evocative character of sweetness and mystery, the fig transforms into a character in its own right in the imaginary worlds of children's stories. The fig becomes a living symbol of sweetness, adventure and lessons in tales and legends intended for curious young minds.

The Fruits of Enchantment

In children's stories, the fig is often described as a fruit with magical powers. Eating a fig can trigger extraordinary events or reveal hidden secrets. This ability to transform reality adds a bewitching dimension to the characters' adventures, transporting young readers into a world of wonder and mystery.

The Mysterious Fig

The fig, with its velvety skin and varied flavors, often becomes a mysterious element in children's stories. Characters may be tasked with finding a rare fig to solve a problem or complete a quest. This quest to unearth the precious fruit adds a touch of suspense and excitement to the story, captivating the readers' imagination.

Hidden Teachings

The fig, with its sweet and sometimes enigmatic nature, is used in stories to teach important lessons. It can symbolize the reward of patience or the discovery of hidden truth. Characters who learn to grow a fig tree and harvest its fruit develop virtues such as perseverance and careful observation.

Adventure in Fruit Land

In some tales, fig trees become portals to magical worlds or distant lands. Children who enter a fig tree can discover enchanted kingdoms, meet fantastical creatures or experience extraordinary adventures. This theme of traveling through a fig tree adds a dimension of wonder and discovery to the story.

The fig, with its sweet and mysterious character, is a rich source of inspiration in children's tales and legends. As a magical fruit, teacher or element of travel, the fig adds a touch of charm and enchantment to these tales. She teaches young readers lessons in patience, courage and adventure, while stimulating their imagination and curiosity. In the fictional worlds of children's stories, the fig becomes a precious ally, bringing a touch of nature and magic to each turned page.

Chapter 54: Fig: Natural Treasure of Antioxidants and Nutrients

The fig, a delicious and seductive fruit, is much more than just a sweet treat. It's a treasure trove of health benefits, rich in antioxidants and essential nutrients. Beyond its sweet flavor, the fig offers a range of natural compounds that nourish the body and protect against the harmful effects of free radicals. the fig reveals itself as a valuable source of antioxidants and nutrients, thus contributing to our well-being and vitality.

Antioxidants: Guardians of Health

Antioxidants are compounds that help protect the body against oxidative damage caused by free radicals. These free radicals are unstable molecules generated by various sources, including pollution, stress and radiation. Antioxidants neutralize these free radicals, helping to reduce the risk of chronic diseases, premature aging and cellular dysfunction.

A Range of Natural Antioxidants

Fig is a storehouse of natural antioxidants, including flavonoids, anthocyanins and carotenoids. These protective plant compounds provide a strong defense against oxidative damage. Anthocyanins, for example, are responsible for the vibrant color of some figs and are known for their anti-inflammatory and antioxidant properties.

Essential Nutrients

In addition to its antioxidant properties, figs are rich in essential nutrients. It is an excellent source of dietary fiber, which contributes to bowel regularity, satiety and overall digestive health. Figs also contain minerals such as potassium, magnesium and calcium, which are crucial for heart, muscle and bone function.

Vitamins for Vitality

Figs are also a source of health-beneficial vitamins. They contain B vitamins, notably vitamin B6, which plays an important role in protein metabolism and the regulation of neurotransmitters. Vitamin K found in figs is essential for blood clotting and bone health.

Overall Health Benefits

Regular consumption of figs and their rich nutrients is linked to a variety of health benefits. It can help maintain heart health by regulating blood pressure and reducing the risk of cardiovascular disease. The fiber in figs supports weight control by promoting satiety and stabilizing blood sugar levels. The antioxidants found in figs are also associated with a reduced risk of certain chronic diseases, including cancer and neurodegenerative disorders.

Fig, much more than just a sweet fruit, is a generous source of antioxidants and nutrients that support health and well-being. Its combination of delicious flavor and health benefits makes it a wise choice for a balanced diet. Whether in the form of fresh figs, dried figs or as a culinary ingredient, the fig makes a valuable contribution to our quest for a healthy and energetic life.

Chapter 55: Fig Roots: Traditional Medicine with Deep Roots

For centuries, plants have been valuable allies of traditional medicine, providing natural remedies to heal and relieve a variety of ailments. Among these beneficial plants, fig roots have played a significant role in various cultures around the world. We will explore the use of fig roots in traditional medicine, revealing a captivating chapter in the relationship between man and nature in the search for healing.

Fig Roots: Anchored in Tradition

Fig roots, rich in phytochemicals and nutrients, have been used in different medical traditions for their healing properties. As the fundamental elements of the plant, roots reflect the energy of the earth and are often associated with qualities of grounding and stability.

Ayurvedic Medicine: Balance and Harmony

In Ayurveda, India's ancient medical system, fig roots are used to balance the doshas (vital forces) and treat a range of health conditions. The anti-inflammatory properties and

The astringent roots are used to relieve gastrointestinal disorders, inflammation and even skin infections.

Traditional Chinese Medicine: Harmonization of Vital Energy

In China, fig roots are valued for their ability to soothe internal heat and strengthen the digestive system. In traditional Chinese medicine, they are often used to balance the energy of the liver and spleen, promoting better digestion and internal harmony.

Mediterranean Traditions: Natural Remedies

Mediterranean regions have also used fig roots for medicinal purposes. These roots are known for their diuretic properties and their beneficial effects on digestion. They have been used to relieve gastrointestinal disorders, stomach aches and discomfort related to the urinary tract.

Phytochemical Properties: Keys to Healing

Fig roots contain various bioactive compounds, including tannins, flavonoids and polysaccharides. These compounds give the roots their anti-inflammatory, antioxidant and antimicrobial properties. Tannins, for example, can help reduce inflammation and protect tissues from oxidative damage.

Ancient Wisdom and Natural Resources

The use of fig roots in traditional medicine is a testament to the ancient wisdom of cultures who learned to harness natural resources for their well-being. Although modern medical practices have evolved, the value of herbal remedies continues to be recognized and studied.

Fig tree roots, steeped in medicinal traditions across the world, offer a fascinating insight into how plants have been used to heal and soothe for generations. The qualities

The healing properties of these roots have been harnessed in varied medical systems, reflecting a deep understanding of the connection between nature and human health. Although traditional medicine has been modified by time, fig roots remain a vibrant testament to the power of natural resources to support healing and balance.

Chapter 56: Royal Elegance: Fig Trees in the Royal and Imperial Gardens

Royal and imperial gardens have always been reflections of the grandeur, aesthetics and refinement of the societies that created them. Among the many elements that adorn these sumptuous gardens, fig trees occupy a special place. Their majestic presence, lush leaves and delectable fruits add a touch of natural elegance to these havens of beauty and serenity. Let's see the enchanting world of fig trees in the royal and imperial gardens, where nature blends harmoniously with human grandeur.

Fig Trees: Plant Jewels of the Royal Courts

Fig trees have long been valued for their ornamental beauty and bounty of succulent fruit. In the royal and imperial gardens, these majestic trees were carefully cultivated to add a touch of lushness to the royal surroundings. Their deep green leaves and graceful forms create a natural backdrop for palaces and grand homes.

The Symbolism of Fertility and Abundance

Fig trees, with their ability to produce fruit abundantly, have often been associated with symbols of fertility, prosperity and abundance. In the gardens of kings and emperors, fig trees express the richness and generosity of nature, while reinforcing the image of royalty as protector and nurturer of its people.

The Intimacy of the Secret Gardens

Fig trees, with their dense leaves and wide wingspan, have also been used to create intimate and shaded spaces in the royal gardens. The avenues lined with fig trees provide peaceful retreats where monarchs could escape the gaze of the outside world and meditate in a serene setting.

Rare and Exotic Varieties

In the royal gardens, the search for exclusivity and exoticism was a distinctive feature. Thus, rare and special varieties of fig trees from various parts of the world were often grown for their unique charm. These exotic fig trees added an international dimension to the gardens, reflecting the cultural influences and international connections of the royal and imperial courts.

The Enduring Legacy

Many royal and imperial gardens still remain today, silent witnesses to past history and grandeur. The fig trees, some of which have survived through the centuries, continue to embody the very essence of these prestigious gardens. They remind modern visitors of timeless elegance and the relationship between nature and nobility.

The fig trees in the royal and imperial gardens transcend time, adding a touch of natural majesty to the sumptuous courtyards of yesteryear. Their graceful silhouettes, lush leaves and succulent fruits evoke the harmonious alliance between royalty and nature. By creating spaces of beauty and tranquility, fig trees have sculpted a plant heritage that continues to captivate and enchant, testifying to the union between the human kingdom and natural splendor.

Chapter 57: The Sweetness of Nature: Fig in Homemade Natural Cosmetics

In our quest for natural and authentic beauty, we have always found refuge in the treasures that nature offers us. Among these treasures, the fig stands out for its captivating sweetness and its benefits for the skin. More and more people are turning to natural, homemade cosmetics to take care of their skin, and fig plays a central role in this process. The fig, this sweet and precious fruit, can be transformed into natural beauty treatments that nourish, revitalize and enhance the skin.

The Charm of Fig in Natural Cosmetics

The fig, rich in antioxidants, vitamins and nutrients, has qualities that make it valuable for the skin. By incorporating fig into homemade cosmetic products, we take advantage of its hydrating, softening and regenerating properties. These qualities make it an ideal ingredient for masks, lotions and scrubs.

Natural Hydration

Fig pulp, full of water, is a powerful natural moisturizer for the skin. By incorporating it into homemade masks and moisturizers, you're giving your skin a dose of essential moisture, helping to maintain its elasticity and natural glow.

Gentle and Effective Exfoliation

The small seeds present in the fig are perfect for gently exfoliating the skin, removing dead cells and revealing a brighter complexion. By using them in homemade scrubs, you get a gentle, non-aggressive exfoliation, leaving skin soft and renewed.

Skin Regeneration

The antioxidants found in fig help protect the skin against oxidative damage caused by free radicals. By creating fig-based masks or serums, you encourage cell regeneration and help delay the signs of aging.

Fig-Based Beauty Recipes

1. **Fig Moisturizing Mask**: Mix fig pulp with natural yogurt and honey. Apply this mask to the face and leave on for 15 minutes before rinsing.

2. **Fig Exfoliating Scrub**. Mix figs with oats and honey to create a gentle scrub. Gently massage onto face in circular motions, then rinse.

3. **Revitalizing Tonic Lotion with Fig**: Steep fig leaves in hot water, let cool and use as a revitalizing skin toner.

4. **Regenerating Fig Serum**: Mix argan oil with prickly pear oil to create a nourishing serum to apply in the evening.

Fig, sweet and delicious, fits harmoniously into the world of natural homemade cosmetics. By using its hydrating, regenerating and exfoliating benefits, you can create beauty treatments that nourish and enhance your skin. By opting for natural fig-based cosmetics, you experience the power of nature while taking care of your natural beauty.

Chapter 58:

The Fig in Asian Cuisine: An Exquisite Taste Journey

Asian cuisine is renowned for its richness, diversity and aromatic complexity. At the heart of this exceptional gastronomy, the fig has managed to carve out a unique place for itself. In Asia, this sweet and fragrant fruit is used creatively in a variety of dishes, from desserts to main courses, adding an exotic and delectable touch to regional cuisine. In this essay, we'll explore how the fig fits seamlessly into Asian culinary culture, bringing its unique flavors and natural sweetness to an already captivating gastronomic palette.

A Sublime Marriage of Flavors

The fig, with its combination of sweetness and rich nutrients, blends harmoniously into the Asian flavor palette. In many Asian cuisines, the notion of balance of flavors is paramount, and the fig, with its sweet profile, adds a note of subtle sweetness that complements the other complex flavors in the cuisine.

In the Main Dishes

In Asia, the fig is used in main dishes to add a sweet and sour touch. Fig-based sauces are often combined with grilled meats, creating a contrast of flavors that stimulates the taste buds.

In Salads and Light Dishes

Fresh fig, with its juicy and crunchy texture, is a great addition to salads and light dishes. It brings a welcome freshness and a sweet dimension to vegetables and aromatic herbs.

In Exquisite Desserts

Asian desserts are often masterpieces of flavors and textures. The fig finds its place in these sweet creations, whether candied figs in pastries, fresh figs in exotic fruit salads or dried figs in traditional sweets.

Cultural and Symbolic Role

In Asia, the fig is sometimes linked to beliefs and cultural symbolism. In some cultures, the fig is considered a symbol of prosperity, abundance and longevity, making it a popular ingredient during special celebrations and holidays.

Culinary Innovation and Fusion

Contemporary Asian cuisine has also seen bold experiments with figs. Innovative chefs are integrating this fruit into fusion dishes, combining Asian culinary traditions with international influences to create unique taste experiences.

The fig, with its sweet flavor and versatility, finds its place at the heart of Asian cuisine. From main courses to desserts, from sweet flavors to sweet and sour notes, the fig fits harmoniously into the Asian culinary palette, adding a touch of exoticism and elegance to dishes already rich in flavor .

By celebrating the fig in their cuisine, Asian cooks pay homage to the diversity and richness of ingredients that nature offers, while providing gourmets with an unforgettable taste experience.

Chapter 59: Fig Trees in Historical Travel Stories: Between Discovery and Wonder

Historical travel stories provide windows into ancient and exotic worlds, where the unknown and wonder abound. Among the wonders that have captivated the imagination of explorers, fig trees occupy pride of place. These majestic trees, symbols of the exoticism and wealth of distant lands, have been immortalized in the writings of daring travelers. Here we will embark on a journey through historical accounts to discover how fig trees fueled the curiosity, admiration and inspiration of traveling explorers.

The Exoticism of Outlands

For historical explorers, fig trees were often synonymous with exoticism and extraordinary discoveries. In their stories, these majestic trees were often described with a fascination mixed with mystery, evoking distant lands that seemed to belong to another world.

Fig Trees in Biblical and Mythological Stories

Fig trees also feature prominently in biblical and mythological stories, adding a sacred dimension to their reputation. From the tree of knowledge in the Garden of Eden to the fig tree under which Buddha achieved enlightenment, fig trees were often linked to moments of revelation and transcendence.

Fig Trees as Landmarks

In their stories, travelers often used fig trees as landmarks to navigate unknown lands. Large fig trees served as natural beacons, providing explorers with orientation and a visual reference to navigate foreign environments.

Wonder at the Generosity of Nature

Fig trees, with their ability to produce abundant fruit, were a source of wonder to travelers. They were fascinated by the generosity of these trees, which offered a multitude of sweet and nourishing fruits. This generosity was often seen as a gift from nature in new lands.

Fig Trees in the Writings of Famous Explorers

Explorers such as Marco Polo and Ibn Battuta mentioned fig trees in their travel accounts, testifying to the impact of these trees on travelers of different eras and cultures. Fig trees were often described as essential parts of the environment and way of life of local people.

Fig Trees as Catalysts of Cultural Exchange

Fig trees, found in various regions of the world, have often acted as cultural bridges between explorers and indigenous populations. They served as symbols of sharing and togetherness, allowing travelers to bond with locals and learn about their customs and traditions.

Fig trees have left their mark on historical travel stories, adding a note of exoticism, admiration and discovery to these fascinating tales. Their lush leaves, abundant fruits and majestic presence captured the imagination of explorers, inspiring fiery and captivating descriptions. Throughout the pages of travel stories, fig trees remind us that human curiosity and exploration have always been guided by the magic of nature and the wonder at its treasures.

Chapter 60: The Art of Fig Wine: When Hedonism Unites with Nature

The world of wine is a universe rich in traditions, know-how and passion. Beyond grapes, some daring vineyards have explored new horizons by using a very different fruit to create unique nectars: the fig. The art of fig wine is a marriage between viticultural craftsmanship and the natural sweetness of this fruit. In this essay, we will explore the fascinating world of fig wine, where the creativity of winemakers combines with the generosity of nature to produce exquisite beverages that arouse wonder and taste pleasure.

When Fig Meets Wine

The fig, with its sweet flavor and fleshy texture, provides an ideal backdrop for creating aromatic and complex wines. Wineries venturing into making fig wine have learned to play with the intricacies of fig varieties, combining different types of fruit to achieve unique flavor profiles.

The Fig Winemaking Process

Fig winemaking requires a specific process that takes advantage of its unique qualities. Figs are often harvested at their peak of ripeness, then pressed to extract their sweet juice. This juice is then fermented, sometimes with the addition of special yeasts to achieve more complex flavors and aromas.

Amazing Flavor Profiles

Fig wine offers a range of flavor profiles from sweet to dry, fruity to complex. Figs add a touch of natural sweetness and fruity notes to the wine's taste palette, creating subtle balances and harmonious contrasts.

A Gourmet Experience

Fig wine is not just limited to tasting. It can be enjoyed at different stages of a meal, as an aperitif or as an accompaniment to dishes. Its natural sweetness makes it an ideal choice for pairing with cheeses, spicy dishes and even desserts.

Fig Wine as Cultural Heritage

In some regions, fig wine making is steeped in history and culture. These wines can be seen as living testimonies of the deep connection between communities and their natural environment. They reflect the ingenuity of previous generations who knew how to transform local resources into gastronomic treasures.

Innovation and Experimentation

The art of fig wine continues to evolve as winemakers explore new methods, new fruit blends and new aromas. Sommeliers and wine lovers are seduced by the originality of these beverages which defy conventions and awaken the taste buds.

Fig wine is a tribute to the union of human creativity and the generosity of nature. Wineries that engage in this subtle art reveal a deep understanding of flavors and transformation processes. The results are exquisite wines that embody the spirit of experimentation and taste exploration. The art of fig wine reminds us that the world of wine is a blank canvas where each fruit, including the fig, can add its unique touch to create unforgettable sensory experiences.

Chapter 61: Fig Trees in Marriage and Birth Traditions: Symbols of Fertility and New Beginnings

Wedding and birth traditions are steeped in symbols and rituals that mark crucial moments in life. Among these symbols, fig trees stand out for their association with fertility, growth and rebirth. Fig trees have found their place in customs surrounding weddings and births, adding a dimension of prosperity and renewal to these celebratory times.

The Fig Tree: Symbol of Fertility

Since ancient times, fig trees have been seen as symbols of fertility and procreation. Their capacity à producing an abundance of sweet fruit has often been associated with the promise of offspring and growth. In wedding and birth traditions, fig trees embody the hope of new lives and new generations.

Fig Trees in Wedding Ceremonies

In some cultures, fig trees are present during wedding ceremonies, whether as decorative elements or symbolic gifts. Figs, with their rounded, fleshy shape, evoke the promise of a fruitful union and the expansion of the family. In some traditions, the bride and groom consume figs or drink fig wine to seal their commitment and their wish for fertility.

Birth and Growth Rituals

During births, fig trees symbolize growth and the start of a new life. Fig saplings can be planted to celebrate the birth of a child, embodying the idea that the child will grow just like the tree. This tradition is a way of wishing the child a long, prosperous and fulfilling life.

Fig Trees and Positive Superstitions

In some cultures, fig trees are surrounded by positive superstitions regarding marriages and births. Figs are considered good luck charms, believed to bring luck, abundance and good vibes to newlyweds and newborns.

The Fig Tree: A Witness to Life in Evolution

The longevity and growth of fig trees make them silent witnesses to the evolution of life and generations. The fig trees planted at weddings and births grow over time, reminding families of the moments of joy and hope that marked their stories.

Renewal and Hope

Fig trees embody the eternal cycle of life, growth and renewal. They remind those celebrating weddings and births that life is constantly changing, offering new opportunities and new promises.

Fig trees play a significant role in wedding and birth traditions, bringing a dimension of fertility, growth and renewal to these special times. Their presence reminds participants that life is a journey full of promise, celebration and hope for the future. Fig trees symbolize the eternal beauty of nature and how it fits harmoniously into the significant moments of human life.

Chapter 62: The Elegant Influence of the Fig Tree on Architecture and Design: A Natural Fusion and Creative

Architecture and design are ways of expressing human creativity and shaping the built environment around us. In this quest for beauty and harmony, nature plays a fundamental role. Among the natural elements that have influenced architecture and design, the fig tree stands out for its majestic presence and intimate connection with the environment. The fig tree, with its elegant silhouette and ability to transform space, has inspired and shaped architecture and design throughout the ages.

Organic and Integrated Architecture

The fig tree, with its lush foliage and calming shade, has often served as a model for the organic integration of nature into architecture. Buildings designed with open interior spaces and courtyards recall the natural shade of fig trees, creating welcoming and refreshing environments.

The Elegance of the Silhouette

The sinuous lines of fig trees, with their intertwining branches and delicate leaves, have inspired architectural elements such as vaults, arcades and decorative designs. These organic forms add a touch of elegance and fluidity to built spaces, creating a feeling of movement and softness.

Shadow and light

Fig trees, with their dense foliage, have an impact on the natural lighting of spaces. Architects and designers have learned to play with the effects of light and shadow produced by fig trees to create unique and evocative atmospheres.

Incorporation of Nature

Outdoor spaces designed around fig trees provide peaceful retreats where individuals can connect with nature. The gardens, courtyards and green spaces surrounding the fig trees offer an oasis of calm amidst the hustle and bustle of the city.

The Art of Furniture and Design Elements

The fig tree also inspires furniture design. The graceful curves of the branches have been reinterpreted in furniture pieces, sculptures and decorative elements, creating a symbiosis between natural form and utilitarian function.

A Bridge Between Past and Present

The use of the fig tree in architecture and design creates a link between tradition and modernity. Elements inspired by fig trees recall the wisdom and elegance of nature, while incorporating contemporary materials and technologies for a distinctly modern approach.

The influence of the fig tree on architecture and design illustrates how nature can serve as infinite inspiration for human creativity. The majestic presence and organic beauty of fig trees have shaped spaces that invite relaxation, contemplation and connection with nature. Architecture and design that incorporates the essence of the fig tree is a celebration of the harmony between humans and nature, uniting functionality and aesthetics in an elegant and timeless dance.

Chapter 63: Dried Figs: A Journey through History, Preparation and Use

Dried figs, these sweet and flavor-packed delights, have a history dating back to ancient times. Sourced from the fig tree, these dried fruits have evolved over the centuries to become a popular snack and versatile ingredient in the kitchen.

Let's dive into the history of dried figs, explore preparation methods, and discover the multiple uses of these fruity gems. A Historical Heritage

Dried figs are not only a tasty snack, they are also steeped in history. From ancient civilizations to the Mediterranean kingdoms, dried figs have been used as a source of nutrition, sweetness and preservation.

The Traditional Drying Method

The process of drying figs is relatively simple, but it requires time and patience. Fresh figs are washed, cut and placed in the sun or in a dry place to dry naturally. This process helps retain nutrients and flavors while reducing water content.

Culinary and Gastronomic Uses

Dried figs are versatile ingredients in cooking. They can be eaten as is as an energy snack, but they also lend themselves to a multitude of recipes. From sweet to savory cuisine, dried figs add a sweet and aromatic touch.

Sweet Treats

In baked goods and desserts, dried figs are used to add natural sweetness and depth of flavor. They can be incorporated into cakes, tarts, jams and even chocolates, creating rich and nuanced taste sensations.

Ingenious Savory Combinations

Dried figs also go wonderfully with savory dishes. They can be used in salads to add a sweet and crunchy note, in meat dishes to provide a sweet-salty contrast, or in cheese platters to balance flavors.

Nutritional and Health Benefits

In addition to their taste qualities, dried figs are rich in fiber, minerals and antioxidants. They are also a natural source of sugar, making them a healthier alternative to industrial sweets.

Cultural Heritage and Symbol of Generosity

In some cultures, dried figs are associated with generosity and sharing. They were traditionally given to travelers and guests as a sign of warm welcome and hospitality.

Dried figs are more than just a treat. They embody a millennia-old history of using and transforming natural resources to meet the taste and nutritional needs of humanity. From their traditional drying to their integration into modern cuisine, dried figs are an example of how nature can be tamed to deliver lasting taste pleasures and nutritional benefits.

Chapter 64: The Fig and Indigenous Spirituality: An Ancient Connection to the Sacred Land

Indigenous spirituality is rooted in a deep relationship with nature, the elements and the cycles of life. In this worldview, every natural element carries meaning and spiritual connection. Among these elements, the fig holds a special place as a symbol of harmony between humans and the earth.

The Fig: A Gift from Mother Earth

For many indigenous cultures, the fig is considered a sacred gift from Mother Earth, an expression of her abundance and generosity. It is often seen as a symbol of fertility and renewal, representing the cycles of life, death and rebirth.

The Fig Tree as a Place of Spiritual Meeting

Fig trees are often chosen as places of spiritual meeting and celebration in indigenous cultures. Their lush leaves provide shade and shelter, creating a space for meditation, sacred stories and rituals.

The Cycle of Life and Death

The fig is often interpreted as a reminder of the eternal cycle of life and death. Fig trees, which bear fruit and lose leaves throughout the year, symbolize the duality of the forces of nature and the notion of constant transformation.

The Interconnection of All Things

In indigenous beliefs, each element of nature is interconnected, forming a complex network of spiritual energy. The fig is seen as an integral part of this web, connecting individuals to the spirits of earth, water, sky and all living things.

Rituals and Ceremonies

Fig trees often play a central role in indigenous rituals and ceremonies. They can be the place for prayers, offerings and songs, creating a sacred space where communities come together to honor ancestors, spirits and nature.

Knowledge Transmission

The fig is also associated with the transmission of ancestral knowledge. Fig trees, rooted in the ground for generations, are considered guardians of wisdom and traditions passed down from generation to generation.

A Spirituality in Harmony with Nature

Indigenous spirituality, imbued with respect for the Earth and its gifts, finds a deep echo in the fig. This fruit symbolizes the way indigenous people live in harmony with nature, honoring its cycles and resources.

The fig embodies indigenous spirituality by celebrating the deep connection between humans and the earth. As a symbol of abundance, the cycle of life and rebirth, the fig speaks to the ancient wisdom and deep spirituality of indigenous people. She reminds us that nature is an invaluable source of spiritual reflection, healing and intimate connection with the universe around us.

Chapter 65: Fig Leaves in Food: A Hidden Treasure of Flavors and Benefits

When we think of fig, it's often the sweet, fleshy fruit that comes to mind. However, in many cultures around the world, the leaves of the fig tree are also revered for their culinary uses and health benefits. These delicate and versatile leaves have long been used to add a unique flavor boost to dishes and for their beneficial properties.

Leaves with Subtle Flavors

Fig leaves have a subtle, slightly herbaceous flavor that adds a unique dimension to dishes. Fresh or dried, they can be used to infuse delicate aromas into various culinary preparations.

Infusion and Perfume

Fig leaves are often used to infuse liquids such as water, milk or oil. This infusion can be used to flavor sauces, stews, teas and desserts, adding a touch fresh and aromatic in nature.

Packaging and Cooking

In some cultures, fig leaves are used as a natural wrapper for cooking food. Food is wrapped in the leaves before being cooked, which adds a delicate flavor and distinctive aroma to the dishes.

A Mediterranean Tradition

Mediterranean cuisine, renowned for its freshness and flavors, also makes use of fig leaves. They are often used to wrap cheeses, stuffed vegetables, fish and even grilled meats.

Health Benefits

Fig leaves are rich in antioxidant compounds, vitamins and minerals. They are also known for their anti-inflammatory properties and their ability to help regulate blood sugar levels.

An Ancient and World History

The use of fig leaves in food dates back to ancient times and is present in many cultures. From traditional Mediterranean dishes to Asian preparations, fig leaves embody a diverse and storied culinary heritage.

Know-How and Cultural Transmission

The use of fig leaves in food is often passed down from generation to generation. It is an integral part of the food culture of many communities, connecting the past to the present through family and traditional recipes.

Creative Exploration

Modern chefs and cooks are also exploring the creative use of fig leaves in contemporary cuisine. Their unique flavor can be incorporated into a variety of dishes, from starter to dessert.

Fig leaves are a little-known treasure in the culinary world, offering delicate flavors and health benefits. Their use in food demonstrates human ingenuity to exploit natural resources for taste and nutritional purposes. By embracing the richness of the fig in all its forms, we discover a palette of flavors and possibilities that enrich our culinary experience and strengthen our connection with nature.

Chapter 66: Fig Trees in Zen Gardens and Meditative Spaces: Source of Serenity and Spiritual Connection

Zen gardens and meditative spaces are havens of tranquility and contemplation, designed to soothe the mind and nourish the soul. At the heart of these harmonious environments, fig trees play a vital role as natural elements that inspire meditation, reflection and spiritual connection.

Natural Symbiosis

Fig trees, with their lush foliage and organic shapes, fit harmoniously into Zen gardens and meditative spaces. Their gentle, calming presence strengthens the connection between man and nature, facilitating a state of calm and serenity.

The Benevolent Shadow

Fig trees provide cooling and protective shade, creating spaces where individuals can retreat from the hustle and bustle outside to find refuge in their inner world. Under the caring shade of fig trees, meditation practitioners can focus more easily on their breathing and presence.

Contemplative Meditation

Fig trees are often integrated into spaces that encourage contemplative meditation. Their organic shapes and simple beauty invite observation and reflection, helping meditators connect with the present moment and find inner balance.

Symbol of Regeneration

The fig tree, with its cycle of growth, fruiting and leaf loss, symbolizes regeneration and renewal. In Zen gardens, it recalls the cyclical nature of life and encourages one to embrace change as an opportunity for spiritual evolution.

Reconnecting to Earth

Fig trees anchor meditative spaces in terrestrial reality, inviting reconnection with the earth and its elements. By touching the leaves or observing the trunk, meditators are brought back to the present moment and to their own existence within the cosmos.

Support for Creativity

Fig trees also inspire artists and creators who frequent Zen gardens. Their shape, texture and energy fuel the imagination and encourage artistic expression in all its forms.

The Unification of Body and Spirit

Fig trees in meditative spaces embody the idea of unity between body and mind. By creating an environment conducive to inner exploration, they encourage harmony between the physical and spiritual aspects of the individual. Fig trees, with their gentle presence and deep meanings, are fundamental elements of Zen gardens and meditative spaces. They bring a touch of sacred nature, promote contemplation and nourish spiritual connection. By creating a bridge between man and the earth, fig trees guide seekers of inner peace towards a profound experience of serenity, reflection and awakening.

Chapter 67: Figs in Alternative Medicine Practices: Between Tradition and Health Holistic

Figs, for millennia, have been appreciated for their sweet flavors and nutritional benefits. However, their value is not limited to cooking. In the field of alternative medicine, figs have also found their place as key ingredients in various practices aimed at promoting holistic health. Figs are used in alternative medicine practices.

Alternative Medicine and Holistics

Alternative medicine focuses on balancing the body, mind and soul to achieve optimal health. Rather than simply treating symptoms, it seeks to treat the person as a whole. Figs, rich in nutrients and beneficial properties, fit naturally into this approach.

Digestion and Detoxification

Figs are known for their high fiber content, which promotes healthy digestion by regulating intestinal transit. In alternative medicine practices, figs are used to stimulate the digestive system and help remove toxins from the body.

Antioxidant Power

Figs, rich in antioxidants such as polyphenols, may help reduce free radical damage. In holistic approaches, figs are integrated to support cellular health and prevent diseases linked to oxidative stress.

Energy Balance

In some practices, figs are associated with specific energetic properties. Their mild flavor is considered soothing to the nervous system, helping to balance energy and promote a feeling of calm.

Weight Management

Figs, being a naturally sweet and fiber-rich snack, are sometimes used in weight management approaches. They can help reduce cravings and keep blood sugar levels stable.

Anchored in Nature

Figs are rooted in the earth and are harvested from a tree, connecting them to nature. In alternative medicine practices, this natural anchor is valued to promote connection with the Earth and restore energy balance.

Cultural Transmission

The use of figs in alternative medicine often reflects cultural traditions passed down from generation to generation. Fig remedies are an integral part of the medical heritage of various communities.

Creative Integration

Figs can be consumed in different ways in alternative medicine practices: fresh, dried, in decoction or even in tincture. Their versatility allows practitioners to customize approaches based on individual needs.

Figs, with their nutrients, antioxidants and natural symbolism, have naturally fit into alternative medicine practices. They embody the notion of holistic health by nourishing both body and mind. Whether by regulating digestion, boosting immunity, or promoting energy balance, figs are an example of how natural resources can be used to support overall health.

Chapter 68: The Poetic and Melodic Echo of the Fig: Between Verses and Folk Songs

The fig, this sweet and fleshy fruit, has long inspired poets and musicians throughout the centuries. Its evocative image, rich flavors and deep symbolism make it a popular subject in poetry and folk songs. In this chapter, we will delve into the poetic and melodic world of the fig, exploring how it has been celebrated and immortalized through the verses and melodies of various cultures.

The Praise of the Senses

The fig awakens the senses with its distinctive aromas and textures. In poetry, she is often described with a range of metaphors that explore her sweetness, tenderness and sensuality. Poets unified the senses of taste, smell and touch to capture the richness of the fig experience.

The Symbolic Figurative

The fig goes beyond its literal nature to become a powerful symbol in poetry. It can represent fertility, nostalgia, transformation or even the human soul. Poets skillfully use the fig to explore deep themes of life and human nature.

The Nature Alliance

The fig is often rooted in its natural environment, surrounded by other elements of nature like trees, rivers and the seasons. This harmonious integration into the natural world adds a dimension of connection to the poetry, strengthening the bond between man and the earth.

The Fig as a Metaphor of Life

The fig, which goes through a cycle of growth, maturity and decline, is used as a metaphor for human life. Its transformation from flower to fruit is likened to the evolution of the human being, making the fig a source of inspiration for poets who meditate on the human condition.

Hidden Stories and Tales

In folk songs, the fig is sometimes the protagonist of stories and tales that teach life lessons, evoke laughter or offer social commentary. These stories are an integral part of the cultural fabric and reflect how the fig is woven into the oral histories of various communities.

The Music of Flavors

Folk songs often highlight the pleasures of the table, and the fig often holds a place of honor. The lyrics describe its sweetness and unique flavor, inviting the listener to imagine the taste while listening to it.

Intergenerational Transmission

Songs and poems about the fig are often passed down from generation to generation, forming a living link between past and present. Families gather to sing songs and recite poems that evoke memories of this fruit and the experiences associated with tasting it.

The fig, with its rich symbolism and cultural heritage, has become a frequent theme in poetry and folk songs. Through verses and melodies, the fig transcends its simple nature to become an object of deep reflection and artistic celebration. Its place in creative expression speaks to its status as a treasured part of human culture, immortalized through the rhythms of speech and the harmonies of music.

Chapter 69: Fig Trees in the Legends of Indigenous Peoples: Spiritual Roots and Stories Sacred

Indigenous peoples across the world have forged deep connections with nature, imbuing their culture with sacred stories and legends that celebrate the connection between humans and the environment. Fig trees, rooted in their territories, have often taken a central place in these legends, symbolizing the symbiosis between humans and nourishing earth.

The Fig Trees Keepers of Knowledge

In many indigenous cultures, fig trees are considered guardians of ancestral knowledge. Trees, with their roots buried deep in the earth, are seen as guardians of teachings and traditions passed down from generation to generation.

The Birth of Fig Trees: Creation Myths

Fig trees often feature prominently in indigenous creation myths. They are sometimes seen as gifts from the gods or as divine creations that rooted life on earth.

The Tree of Life and Regeneration

Fig trees are often associated with the idea of the tree of life, symbolizing the continuity of life, regeneration and rebirth. Their ability to give birth to new trees from their roots has reinforced their status as a symbol of renewal.

Healing and Well-Being

In legends, fig trees are often linked to healing and well-being. Their leaves, fruits and roots are revered for their healing properties. Fig trees thus become spiritual allies in the quest for physical and spiritual health.

Sacred Encounters

Fig trees are sometimes referred to as meeting places between the spiritual and earthly worlds. Beneath their branches, indigenous people hold ceremonies, prayers and rituals, creating sacred spaces to communicate with spirits and ancestors.

Transformation Legends

Some legends tell how individuals transformed themselves into fig trees, thus becoming guardians of the earth and protectors of territories. These stories highlight the close bond between man and nature.

Intergenerational Transmission

Legends of fig trees are passed down from generation to generation, embodying the collective memory of indigenous peoples. They are told around the fire, during ceremonies and in moments where ancient wisdom is shared with younger generations.

Cultural Revealers

These legends reveal unique perspectives on spirituality, cosmology and the relationship between human beings and the earth. They offer profound insight into how indigenous people view their place in the universe.

Fig tree legends, central to indigenous cultures, are treasures of wisdom that tell stories of creation, healing, regeneration and spiritual connection. Rooted in respect for the land and ancestors, these stories reflect the spiritual depth of indigenous peoples and testify to the rich connections between man and nature.

Chapter 70: The Art of Fusion: The Fig in Contemporary Molecular Cuisine

Contemporary molecular cuisine, with its bold exploration of the physical and chemical properties of ingredients, has revolutionized the way we experience flavors and textures. Among the many ingredients that have been reinvented in this context, the fig stands out for its aromatic complexity and unique structure.

The Fusion of Flavors

The fig, with its combination of sweet and slightly tart, provides fertile ground for culinary experimentation. In molecular cuisine, chefs use methods like spherification, gelling and foamation to capture and intensify its aromas, creating unique taste explosions.

Creative Texturization

One of the most interesting characteristics of the fig is its texture, which is both tender and crunchy. Molecular cuisine techniques allow chefs to explore these textures in innovative ways, creating unexpected contrasts and transforming the fig into a tactile and sensory experience.

Spherification and Gelation

Fig can be transformed into delicate pearls through spherification, creating flavor capsules that burst in the mouth. Gelling creates gelatinous and melting textures, offering a new dimension to tasting.

Emulsification and Foamation

Fig can be transformed into light and airy mousses through foamation, providing a fascinating taste and visual experience. These mousses allow you to play with the perception of flavors and textures.

The Fig as an Edible Work of Art

In molecular cuisine, presentation plays an essential role. Chefs transform the fig into true edible works of art, combining visual, aromatic and taste elements to create an immersive gastronomic experience.

New Perspectives on Fig

Contemporary molecular cuisine offers new perspectives on the fig, allowing chefs to rethink how it can be used in sweet and savory dishes. Bold combinations with other unexpected ingredients expand the range of culinary possibilities.

The Emergence of New Dishes

The fig, once subjected to molecular cuisine techniques, is transformed into an ingredient for innovative and surprising dishes. From spherical desserts to savory dishes where fig is integrated in surprising ways, the creativity of the chefs pushes the limits of culinary imagination.

A Gastronomic Heritage

The fig has a long history in traditional cuisine, and its incorporation into molecular cuisine as a modern ingredient reinforces its status as a versatile and captivating ingredient.

The fig, with its complex flavor profile and unique texture, has found new life in contemporary molecular cuisine. The innovative techniques of this cuisine make it possible to deconstruct, transform and reinvent this emblematic fruit. The fig thus becomes a blank canvas for chefs who wish to create unique gastronomic experiences, pushing the boundaries of culinary creativity.

Chapter 71: Emerging Crafts: The Use of Fig Roots in Creation

Craftsmanship, fusing creativity and traditional know-how, often finds its inspiration in natural elements. Fig tree roots, long neglected, have recently attracted the interest of artisans across the world. Their tortuous shape and solidity offer unexplored potential for creating unique and durable objects.

Recovered Natural Resources

The use of fig roots in crafts is part of the trend of recovery and sustainable use of natural resources. Rather than leaving these roots unused, artisans transform them into original pieces of craftsmanship, minimizing waste.

Exploration of Natural Forms

Fig tree roots are known for their intricate, organic shapes. Artisans exploit these characteristics by integrating them into the design of furniture, sculptures and decorative objects. Each root is unique, giving the works a touch of authenticity and character.

Functional and Artistic Objects

Fig tree roots are used to create a variety of items, from tables and chairs to lamps and picture frames. Their use in functional objects adds an artistic dimension to everyday life, transforming utilitarian items into pieces of art.

Alliance of Natural and Cultural

The use of fig roots creates a bridge between nature and culture. Artisans respect the original form of the root while integrating it into cultural and aesthetic contexts, giving rise to objects that tell a story of both nature and human creativity.

Sustainability and Authenticity

Crafts based on fig tree roots are part of the quest for more sustainable and authentic consumption. Objects created from natural materials often reflect the values of the artisan and the consumer, emphasizing originality and sustainability.

A Complex Creative Process

Working with fig roots requires specific skills and techniques. Artisans must understand the nature of the material, its strength and its possibilities, which adds a layer of complexity to their creative process.

Cultural Heritage

The use of fig roots in crafts may also be linked to specific cultural traditions. In some communities, these roots have symbolic and spiritual meaning, which reinforces their presence in local crafts.

The use of fig roots in crafts is a testament to how human creativity can transform natural elements into works of art. Crafts based on fig roots celebrate beauty of nature while embodying the know-how and creativity of artisans. These unique objects, at the crossroads of the natural and the artistic, testify to the possible harmony between man and his environment.

Chapter 72: The Mystical Imprint: Fig Trees in Middle Eastern Culture

The Middle East, rich in history and tradition, has forged deep ties with the fig tree, a tree that goes beyond its physical nature to become a symbol of spirituality, sharing and cultural heritage. Fig trees have rooted their roots in the daily life, spirituality and traditions of this region.

The Tree of Life and Spirituality

The fig tree is often considered a tree of life in Middle Eastern culture, embodying the connection between the earth and the divine. The ancient fig trees that have thrived in this region for centuries are revered as guardians of spirituality and wisdom.

Symbol of Sharing and Hospitality

In many Middle Eastern cultures, figs are associated with generosity and hospitality. Fresh and dried figs are often offered to guests as a sign of warm welcome and sharing, creating moments of conviviality and connection.

Traditional Know-How

The processing of fresh figs into jams, fruit jellies and other sweet delights is rooted in the culinary heritage of the Middle East. These preparations are often passed down from generation to generation, thus preserving traditional know-how and the link with the past.

Symbol of Prosperity and Sustainability

Fig trees are also seen as a symbol of prosperity and sustainability. Their ability to thrive in arid environments has often been interpreted as a message of resilience and abundance.

The Tree of Encounters and Stories

Under the shade of fig trees, important meetings take place, stories are told and traditions are shared. These trees become gathering points where generations come together to exchange knowledge, celebrate and connect.

Wisdom and Heritage

Ancient fig trees have a dominant presence in the landscapes of the Middle East. Their deep-rooted roots symbolize the connection with the past, remembering previous generations and passing on their wisdom to new generations.

Incorporation into Art and Literature

Fig trees have often inspired poets, writers and artists in the Middle East. Their image is found in poetry, stories and artistic works, where they often represent metaphors for life, spirituality and beauty.

Rituality and Celebrations

Fig trees are often incorporated into religious rituals and celebrations. Their symbolic role in various spiritual traditions reinforces their sacred status and links them to significant moments in life.

Fig trees have deeply rooted their presence in Middle Eastern culture, becoming symbols of hospitality, spirituality and the connection between humans and nature. Their cultural and spiritual heritage, as well as their role in daily life, testify to their deep place in the social and cultural fabric of the region. Middle Eastern fig trees are keepers of memory, tradition and spiritual reflection, reflecting the cultural richness and diversity of this ancient land.

Chapter 73: The Fig and Food Sustainability: A Responsible Food Alliance

In a constantly changing world, food sustainability has become a major concern for individuals, communities and the planet. Fig trees, with their historical heritage and contribution to food security, play a crucial role in this quest for sustainability. The fig fits into concepts of food sustainability, promoting the preservation of the environment and the well-being of communities.

An Ancient Fruit with a Promising Present

The fig trees, ancient and resilient, have stood the test of time. Their ability to grow in arid environments and their contribution to the human diet for millennia make them valuable allies in promoting food sustainability.

Local Cultures and Resilience

In many regions, fig trees are an integral part of local food systems. Their local cultivation and consumption builds community resilience by reducing dependence on imported foods and preserving culinary traditions.

Low Environmental Impact

Fig trees often require less water and chemical inputs than other crops. Their adaptability to harsh climatic conditions makes them a sustainable choice for areas subject to drought and climatic variations.

Biodiversity and Agroforestry

Fig trees can be integrated into agroforestry systems, promoting biodiversity and soil regeneration. Their deep roots help prevent erosion, while their presence can support other crops and plant species.

Resource Conservation

Processing figs into products such as jams, fruit jellies and dried figs extends their shelf life. This helps reduce food waste and maximize crop utilization.

Local Economy and Economic Sustainability

The cultivation and marketing of figs can boost the local economy, creating jobs and promoting the economic sustainability of farming communities.

Promotion of Food Sovereignty

By focusing on growing and consuming local figs, communities can strengthen their food sovereignty by having increased control over their food supply.

Food Education and Awareness

The fig can serve as a bridge to educate consumers about the importance of choosing sustainable and local foods. By sharing stories about fig trees, we can inspire positive dietary behavior change.

Fig trees illustrate how a natural resource, with its history and ability to thrive in harsh conditions, can be an essential component of food sustainability. By encouraging the cultivation and consumption of local figs, exploring processing methods and integrating fig trees into agroforestry systems, we can create a future where food is abundant, diverse and environmentally friendly. The fig, as a symbol of food sustainability, reminds us that the choices we make today impact future generations and the health of the planet.

Chapter 74: Natural Balance: Fig Trees and Chinese Healing Practices

For millennia, traditional Chinese medicine has relied on a holistic view of health, emphasizing the harmony between body, mind and nature. Fig trees, with their properties nutritional and medicinal, have a significant place in these practices. Fig trees are integrated into Chinese healing practices, contributing to the search for balance and well-being.

The Foundations of Traditional Chinese Medicine

Traditional Chinese medicine is based on the concept of Qi (vital energy) and the balance between Yin and Yang. Fig trees, with their balanced nature and health effects, fit harmoniously into this philosophy.

Fig Trees in Chinese Dietetics

In Chinese dietetics, foods are classified based on their thermal properties and their effects on the body. Figs, often considered fresh and slightly cooling, are used to balance internal heat and treat imbalances.

Nourish Yin and Blood

Figs are considered beneficial for Yin (the nourishing, feminine aspect of energy) and Blood (which encompasses vitality and regeneration). They are often recommended to treat dryness, dry cough, menstrual disturbances and other symptoms associated with Yin deficiency.

Strengthening the Spleen and Stomach

In traditional Chinese medicine, the Spleen and Stomach are responsible for digestion and assimilation of nutrients. Figs, with their mild flavor and nourishing nature, are often recommended for strengthening these organs.

Antioxidant and Nutrient Properties

Figs are rich in antioxidants, fiber and minerals, making them valuable for supporting digestive health, reducing inflammation and strengthening the immune system, important elements in traditional Chinese medicine.

Integration into Herbal Formulas

Figs can be used alone or combined with other herbs to create herbal formulas in Chinese medicine. These combinations aim to treat specific conditions by taking into account the complex interactions between ingredients.

Energy Practices and Meditation

Fig trees are sometimes integrated into gardens and environments conducive to meditation, promoting calm and relaxation. Spaces where fig trees thrive are considered conducive to the cultivation of internal energy.

Respect for Balance

The integration of fig trees into Chinese healing practices highlights the importance of respecting the balance between the individual, nature and cosmic forces. Fig trees, with their natural qualities, are considered an extension of this harmony.

Fig trees play an essential role in Chinese healing practices by contributing to harmony of body and mind. Their balancing nature, their nutritional properties and their effects on health make them valuable allies in the quest for holistic well-being. Fig trees, rooted in Chinese healing philosophy and practices, serve as a reminder of the importance of harmony and connection with nature in cultivating a healthy, balanced life.

Chapter 75: The Sweet Craft: The Art of Making Fig Jam

Fig jam making is a craft that dates back centuries, a practice that marries the sweetness of figs with the art of culinary processing. This meticulous and creative process captures the flavor and richness of figs while extending their shelf life.

Ingredient Selection

The first crucial step in making fig jam is selecting the ingredients. From fresh figs, ripe to perfection, to natural sweeteners such as cane sugar or honey, each ingredient contributes to the final flavor and texture of the jam.

Preparation of Figs

The figs are washed, peeled and pitted. Some recipes may retain the skins for a more rustic texture, while others favor a smoother texture. The figs are then cut into pieces for easier cooking and processing.

Cooking and Processing

The cut figs are combined with the chosen sweetener and cooked over low heat. While cooking, figs gradually break down, releasing their sweet flavors and unique character. Some recipes may also include spices like cinnamon, vanilla or ginger to add aromatic complexity.

Reduction and Thickening

As the fig jam cooks, it reduces in volume and thickens. Prolonged cooking allows the ingredients to mix harmoniously and reach the desired consistency. The art of jam making lies in the ability to adjust the cooking to obtain the perfect texture.

Consistency Check

To determine if the jam is ready, you can use the "drop" technique: placing a small amount of jam on a cold plate, observe its consistency by pushing it lightly with a spoon. If the jam gels and does not flow immediately, it is ready.

Potting and Storage

Once the jam reaches the desired consistency, it is carefully poured into sterilized jars.

The potting process requires careful hygiene to ensure long-term preservation. The jars are sealed airtight to preserve the freshness and flavor of the jam.

Tasting and Appreciation

The artisanal fig jam is ready to enjoy once it has cooled and set. It can be enjoyed on bread, toast, cookies or even used as an ingredient in desserts and savory dishes. Each bite is a tribute to the artistic transformation of figs into a delicious treat.

An Art in Perpetual Evolution

The art of making fig jam evolves over time, incorporating modern techniques and innovative ideas. Jam artisans continue to push the boundaries of creativity by experimenting with flavor combinations and varied cooking processes.

Fig jam making is an art that celebrates the beauty and diversity of figs while capturing their essence in a jar. This meticulous and creative craft demonstrates the relationship between nature and cuisine, where figs become a canvas on which jam makers trace unique tastes and aromas. A spoonful of fig jam is much more than a sweet treat - it embodies effort, tradition and a love for the flavors of nature.

Chapter 76: Between Mysticism and Symbolism: The Fig in Esoteric Beliefs

Fig trees, with their majestic presence and mesmerizing fruit, have always captivated the human imagination. Beyond their physical appearance, fig trees have also found their place in esoteric beliefs, where they embody deeper and mystical meanings.

Spiritual Anchoring and Connectivity

In many esoteric beliefs, fig trees are seen as bridges between the material world and the spiritual world. Their roots buried deep in the ground are interpreted as a symbol of anchoring and connection with the energies of the earth.

Tree of Knowledge and Revelation

The fig tree is also often associated with the tree of knowledge in various esoteric traditions. Its symbolism dates back to biblical mythology where the fig tree represents the quest for knowledge and wisdom.

Hidden Wisdom and Mysteries

Fig trees, which bear sweet fruit hidden beneath their leaves, have been seen as guardians of mysteries and hidden wisdom. Figs, hidden from view, are seen as a reminder that deep knowledge can be discovered by those who seek with determination.

Cycles of Life and Death

The fig tree, which goes through cycles of growth and dormancy, reflects the natural cycles of life and death. In esoteric beliefs, this can be interpreted as a reminder of the importance of accepting the cycles of life and appreciating each phase.

Yin-Yang Harmony

Figs, with their fleshy interiors and soft skin, often embody the harmony of Yin and Yang. This duality is seen as a reminder of the necessary balance between opposing forces in the universe.

Divination Tool

In some esoteric practices, figs have been used for divination. The patterns of the seeds in a halved fig can be interpreted as symbols or messages from the cosmos.

Protection and Spiritual Energy

Fig trees have been used as protective amulets in some esoteric traditions, believed to protect against negative energies and attract positive energies.

Spiritual Renewal

The ability of fig trees to regrow after periods of dormancy has been interpreted as a symbol of spiritual renewal and resilience in esoteric beliefs.

Fig trees in esoteric beliefs illustrate how nature can be interpreted as a mirror of the mysteries of the universe and hidden dimensions of reality. Fig trees are not only physical trees, but also portals to deeper thoughts of wisdom, knowledge and connection. In the esoteric world, fig trees serve as keys to unlock doors to spiritual understanding and the mysteries of life.

Chapter 77: Deep Roots: Fig Trees in Traditional African Culture

Fig trees, with their majestic presence and sweet fruit, hold a special place in traditional African culture. Rooted in the beliefs, customs and practices of communities across the continent, fig trees are much more than just a source of food.

Symbols of Protection and Spirituality

Fig trees are often seen as symbols of protection and spirituality in many African cultures. Towering, majestic trees are often considered habitats for spirits and ancestors, serving as links between the spirit world and the earthly world.

Community Gathering and Sacred Spaces

Fig trees are often community gathering places, serving as meeting points where stories are shared, advice is given and ceremonies are performed. These majestic trees may also be associated with sacred spaces where religious and cultural rituals are observed.

Links with Ancestral Heritage

In many African cultures, fig trees are considered guardians of ancestors and collective memory. Ancient fig trees are revered as guardians of stories, traditions and knowledge passed down from generation to generation.

Symbolism of Growth and Renewal

The growth of the fig tree from a small seed to a majestic tree is often interpreted as a symbol of renewal and personal development. Fig trees reflect the natural cycles of life and renewal, inspiring individuals to embrace change and growth.

Medicinal and Magical Practices

Fig trees are sometimes used in traditional African medicine for their medicinal properties. The leaves, fruits and bark are used in the preparation of remedies for various ailments. Fig trees may also play a role in magical practices, promoting healing, protection and communication with spirits.

Fig Trees and Proverbs

Fig trees are often mentioned in African proverbs and expressions, carrying teachings about patience, growth and wisdom. These proverbs reflect the cultural and spiritual importance of fig trees in daily life.

Art and Crafts

Fig trees have also inspired African art and crafts, with sculptures, weavings and art objects often adorned with fig tree designs. These works of art demonstrate the integration of fig trees into creativity and cultural expression.

Fig trees are more than just a part of the African landscape. They are the guardians of tradition, the spiritual shrines and symbols of connection with the land and ancestors. Deeply rooted in traditional African culture, fig trees continue to serve as links between the past and present, providing a living reminder of the importance of cultural and spiritual heritage.

Chapter 78: Purification and Transcendence: The Fig in Purification Rituals

Since the dawn of humanity, fig trees have been revered for their nutritional and medicinal properties, but also for their spiritual potential. In many cultures around the world, fig trees have been incorporated into purification rituals, symbolizing the quest for physical, mental and spiritual purification. We will delve into the depths of purification rituals linked to fig trees, exploring their symbolism and transformative power.

The Symbolism of Purification

Fig trees, with their natural cycle of growth and renewal, have often been seen as symbols of transformation and regeneration. This intrinsic symbolism harmonizes perfectly with the idea of purification, which aims to get rid of impurities to allow new spiritual growth.

Purification of Body and Spirit

In many cultures, fig trees are used to aid in the purification of the body and mind. Figs, rich in fiber and antioxidants, are considered purifying foods, helping to eliminate toxins from the body. Additionally, fig trees often serve as places of meditation and contemplation, allowing individuals to purify their thoughts and emotions.

Purification of Sacred Places

Fig trees are also associated with the purification of sacred places. Their majestic and calming presence is believed to balance energies and purify the spiritual environment. Some rituals include practicing meditation or prayer under a fig tree to reconnect with divine energies and free oneself from the energies negative.

Purification of Negative Energies

Fig trees are sometimes used in rituals aimed at warding off negative energies or protecting against malicious influences. Fig trees are considered protective guardians, helping to repel harmful forces and create a space of purification and security.

Rituals of Transcendence

Fig trees, with their ability to grow and thrive in varied conditions, are often revered for their resilience. In some purification rituals, fig trees are used to symbolize the individual's ability to transcend challenges and free themselves from the shackles of the past.

Meditation and Inner Purification

Fig trees are also associated with meditation and the search for inner purification. Meditating under a fig tree is believed to facilitate concentration and inner peace, thereby allowing the individual to rid themselves of negative thoughts and connect with their deepest self.

In purification rituals, fig trees act as guides toward transcendence and regeneration. Their calm, protective presence inspires individuals to shed emotional burdens, free themselves from toxins in the body and mind, and seek purity and spiritual elevation. Fig trees become sanctuaries of transformation, providing a constant reminder of the power of introspection and purification to achieve holistic growth.

Chapter 79: Between Earth and Sky: Fig Trees in Sacred Architecture

Sacred architecture, an art that marries human creation and the divine, has often integrated natural elements to express the connection between the earthly and the spiritual. Among these elements, the fig trees stand majestically, bringing deep symbolism to these places of devotion.

The Alliance between Nature and Spirituality

Fig trees, with their ability to connect sky and earth through their branches and roots, embody the intimate alliance between nature and spirituality. In sacred architecture, fig trees are often integrated to symbolize this connection, creating spaces where worshipers can feel the divine presence through nature.

Fig Trees as Symbols of Refuge

The broad wingspan of fig trees, their bushy leaves and their outstretched branches have often inspired the idea of refuge and protection. In places of worship, fig trees are sometimes planted to create shaded spaces where worshipers can find spiritual shelter and feel in communion with divine energies.

Symbols of Spiritual Growth

The growth of fig trees from small seed to majestic tree can be interpreted as a symbol of spiritual growth. Places of worship decorated with fig trees remind the faithful of the need to cultivate their own spirituality and progress on their inner path.

Rituals and Celebrations under the Fig Trees

Fig trees have often been silent witnesses to religious rituals and ceremonies. Sacred spaces surrounded by fig trees provide a natural setting for meditation, prayer and celebration practices. The fig trees then become witnesses and invisible participants in the spiritual quest of individuals.

Divine Encounters under the Fig Trees

In many religious traditions, fig trees are mentioned as places of divine encounters. From biblical stories to ancient myths, fig trees have witnessed moments of revelation and exchange between human beings and the divine.

The Energy of the Sacred Tree

Fig trees, often considered sacred trees in some cultures, are believed to be imbued with a special energy. Their presence is believed to foster connection between worshipers and the divine, as well as between human beings and nature.

Balance and Architectural Harmony

Fig trees, with their harmonious shape and calming presence, add a dimension of balance and harmony to sacred architecture. They create a harmonious contrast between hard and soft, solid and living.

Fig trees, with their powerful symbolism and imposing presence, have found a special place in sacred architecture throughout ages and cultures. They are not simply decorative elements, but portals to spirituality, guardians of the connection between man and the divine, and eternal witnesses to the spiritual quests of humanity. In sacred architecture, fig trees become emissaries of nature, facilitating communion between the earthly and the celestial.

Chapter 80: When Flavors Merge: The Fig in Asian Fusion Gastronomy

Fusion gastronomy is a culinary art form that transcends cultural boundaries to create unique taste experiences. In the Asian context, where culinary diversity is rich and varied, the integration of fig into fusion dishes creates unexpected and delicious combinations.

Meeting of Sweet and Spicy Flavors

Fig, with its sweet sweetness and honeyed notes, pairs perfectly with the spicy and umami flavors of Asian cuisine. In fusion gastronomy, figs can be used to balance spicy dishes, adding a touch of sweetness that softens flavor profiles.

Figs in Savory Dishes

In Asian fusion gastronomy, figs can be incorporated into a variety of savory dishes. By for example, they can be added to rice dishes, vegetable salads, or seafood dishes to bring a sweet and juicy dimension that pleasantly surprises the taste buds.

Fig Sauces and Marinades

Figs can be made into delicious sauces and marinades to enhance Asian dishes. Their rich flavors add complex depth to sweet and sour sauces, teriyaki marinades and meat glazes.

Creative Desserts

In Asian fusion gastronomy, figs lend themselves to the creation of innovative desserts. They can be used to top matcha cakes, mochi pancakes or sticky rice balls, adding a sweet and fruity twist to Asian classics.

Creation of Unique Dishes

Adding figs to traditional Asian dishes can create unique fusion dishes. For example, dishes like Peking duck can be reinvented using figs to bring a new dimension of flavor and texture.

Figs and Tea

The marriage of fig and tea is an interesting characteristic of Asian fusion gastronomy. Figs can be infused in hot or cold teas to create refreshing, fragrant drinks that combine the sweet notes of the fig with the aromatic flavors of the tea.

Exploration of Creativity

Asian fusion gastronomy is a blank canvas for culinary creativity. Incorporating fig into this fusion allows chefs to play with bold flavor combinations, while respecting the rich Asian culinary tradition.

The fig in Asian fusion gastronomy embodies the meeting between old and new, sweet and savory, the familiar and the unexpected. It adds a touch of refinement and originality to beloved Asian dishes, while celebrating diversity and culinary creativity. Figs, with their natural charm and palette of flavors, harmonize perfectly with the culinary delights of Asia, providing an unforgettable gastronomic experience for lovers of fusion and taste discoveries.

Chapter 81: Leaves of Life: Traditional Crafts and Fig Leaves

In cultures around the world, traditional crafts demonstrate creativity and the deep connection between humans and nature. Among the natural resources used to create unique works of art, fig leaves occupy a special place. These versatile leaves, with their delicate textures and shades of green, have been transformed into fascinating works of art across generations.

The Origin of Creativity

Fig leaves have been used in traditional crafts since ancient times. Artisans discovered that these leaves, with their unique shape and natural durability, were ideal for creating objects that were both functional and aesthetic.

Weaving and Braiding

Fig leaves are often used to weave or braid baskets, mats and rugs. Their flexible fibers are woven together to form intricate patterns, creating utilitarian works of art that reflect local culture and aesthetics.

Art of Basketry

Basketry is one of the most popular areas of crafts using fig leaves. From bags to hats to baskets, artisans skillfully transform leaves into functional and elegant objects.

Natural Painting

Some cultures use fig leaves as a canvas for painting. Artisans paint patterns and scenes from daily life on the dried leaves, creating unique and ephemeral works of art.

Sculptures and Ornamentation

Fig leaves are also used to create sculptures and ornamentation. Artisans shape the leaves into various shapes, combine them with other natural materials or paint them to create decorations inspired by nature.

Religious and Spiritual Crafts

In some cultures, fig leaves are used to create religious and spiritual objects, such as offerings, icons or devotional items. These creations capture the spirituality and aesthetics of the culture.

Sustainability and Cultural Heritage

The use of fig leaves in traditional crafts is not only aesthetically appealing, but it also contributes to sustainability. By using natural and renewable materials, artisans preserve the environment while perpetuating ancient artisanal practices.

Transmission of Knowledge

Traditional crafts often involve the transmission of knowledge from one generation to the next. The techniques of weaving, braiding and carving fig leaves are passed down from father to son, preserving a rich cultural heritage.

The use of fig leaves in traditional crafts is an ode to human creativity and natural beauty. These sheets, often neglected in the modern tumult, offer artisans the possibility to create unique and timeless works of art. By continuing these traditions, artisans honor nature and preserve their culture while creating objects that transcend time and space.

Chapter 82: Elegant Evolution: Fig Trees in Contemporary Botanical Gardens

Contemporary botanical gardens are living sanctuaries where nature and art come together to create diverse ecosystems and enchanting landscapes. Among the plant treasures that adorn these green havens, fig trees stand out for their beauty, their historical symbolism and their contribution to environmental education. Let's explore the importance of fig trees in contemporary botanical gardens, and how they enrich the visitor experience while promoting conservation and connection with nature.

Ecology and Diversity

Fig trees, with their multiple species and varieties, contribute to the biological diversity of contemporary botanical gardens. From tropical fig trees to cold-hardy varieties, these versatile trees adapt to different climates and regions, creating a rich, ecologically balanced environment.

Aesthetic Attractions

The majestic silhouette of fig trees, with their broad leaves and elegant branches, gives botanical gardens an aesthetic that attracts the eye and soothes the mind. They often serve as focal points in landscaping, adding a visual dimension that charms visitors.

Environmental Education

Fig trees provide valuable opportunities for environmental education. Botanical gardens often use fig trees to illustrate concepts such as symbiosis, pollination, plant diversity and ecological adaptations. They educate visitors about the complex interactions between plants and their environment.

History and Culture

Fig trees carry deep historical and cultural significance in many societies. Contemporary botanical gardens can use these trees to tell cultural stories, explain their use in traditional medicine, and celebrate their place in the collective imagination.

Conservation and Preservation

Contemporary botanical gardens play a crucial role in the conservation of endangered plant species. Fig trees, some vulnerable due to climate change and habitat loss, find refuge in these gardens. Conservation efforts help maintain these species for future generations.

Interaction and Meditation

Fig trees, with their calming shade, provide ideal places for human interaction with nature. Visitors can rest under their branches, meditate or simply enjoy the serene atmosphere they create.

Scientific Research

Contemporary botanical gardens also serve as research centers. Fig trees are subjects of study to understand their growth, reproduction, disease resistance and their role in ecosystems.

Innovation in Design

Fig trees also inspire innovative approaches in the design of contemporary botanical gardens. Their incorporation into vertical structures, hanging gardens or experimental landscape concepts adds a touch of originality to these spaces.

Fig trees, with their timeless charm and ecological richness, are key players in gardens contemporary botanicals. By blending aesthetics, education, preservation and connection with nature, they enrich visitor experiences and inspire a renewed commitment to protecting our natural world. In these havens of beauty and knowledge, fig trees bear witness to the elegant evolution and harmonious continuity between man and nature.

Chapter 83: The Beauty of Nature: Fig and Herbal Cosmetics

In our quest for effective skin care and beauty products, herbal cosmetics have gained popularity for their natural and beneficial properties. Among the botanical gems used in these formulations, fig stands out for its nourishing and regenerating benefits for the skin. Let's observe the harmonious marriage between fig and herbal cosmetics, revealing how this natural collaboration provides a holistic and nature-friendly beauty experience.

Richness in Natural Nutrients

Fig, full of vitamins, antioxidants and minerals, is a valuable source of nutrients for the skin. Fig-based cosmetics capture this natural richness, providing essential nourishment for radiant, glowing skin.

Deep Hydration

The fig is also known for its high water content. Fig-based cosmetics provide deep hydration to the skin, helping to maintain moisture balance and prevent dehydration.

Antioxidants for Protection

The antioxidants in fig, such as polyphenols, help protect the skin from damage caused by free radicals and environmental exposure. Fig-based cosmetics act as a natural barrier to preserve the health and youthfulness of the skin.

Cellular Regeneration

The fig's natural enzymes promote cell regeneration, making it a valuable ally in anti-aging products. Fig-based cosmetics help reduce the appearance of wrinkles and encourage skin to renew itself

Skin Soothing

The anti-inflammatory properties of fig soothe sensitive or irritated skin. Fig products help calm redness and restore the skin's natural balance.

Harmony with Nature

Fig-based cosmetics embody an approach that respects nature. By utilizing the benefits of this plant, cosmetic manufacturers often eliminate the need for harsh chemicals, promoting sustainable, environmentally friendly beauty.

Sensory Experience

Fig-based cosmetics offer a unique sensory experience. Creamy textures, subtle aromas and soothing benefits create a rich and indulgent skincare experience.

Ethical Commitment

By choosing fig-based cosmetics, consumers often support ethical production and sourcing practices. This strengthens the sustainable supply chain and contributes to the preservation of the environment.

Fig, with its range of natural skin benefits, has become a valuable ingredient in the herbal cosmetics industry. By capturing the power of nature in eco-friendly formulations, fig-based cosmetics offer a path to radiant, nourished-from-within beauty. They highlight the growing importance of returning to the roots of nature for fulfilling and holistic beauty.

Chapter 84: Mediterranean Harmony: Figs in Modern Cuisine

Modern Mediterranean cuisine is a celebration of freshness, simplicity and authentic flavors. At the heart of this culinary tradition is the judicious use of local and seasonal ingredients. Among these ingredients, figs shine as a culinary gem, bringing a sweet and lush touch to Mediterranean dishes. Let's show how figs are integrated into modern Mediterranean cuisine, adding a delicious and unexpected dimension to this beloved culinary tradition.

Season and Authenticity

Modern Mediterranean cuisine celebrates the freshness of seasonal ingredients. Figs, when in season, bring a sweet and juicy note to dishes, creating an authentic culinary experience true to natural cycles.

Elegant Starters

Figs are often used to create elegant and sophisticated starters. For example, they can be combined with local cheeses, prosciutto or nuts to create mouth-watering dishes that tantalize the taste buds.

Fresh Salads

Figs add a touch of sweetness to salads, balancing the fresh flavors of the vegetables. They can be combined with ingredients such as spinach, citrus fruits, nuts and goat cheeses to create colorful and tasty salads.

Inventive Main Dishes

In modern Mediterranean cuisine, figs can be used to create inventive main dishes. For example, they can be added to meat sauces, fish dishes or tagines to add a touch of complex sweetness.

Pizzas and Breads

Figs bring a unique dimension to pizzas and Mediterranean flatbreads. They can be used as a garnish with cheese, grilled vegetables and aromatic herbs, creating surprising and delicious flavor combinations.

Gourmet Desserts

Figs find their peak in modern Mediterranean desserts. They can be made into jams, cakes, tarts and pastries to add natural sweetness and an enchanting scent.

Art of Presentation

In modern Mediterranean cuisine, presentation is just as important as the flavors themselves. Figs, with their appealing aesthetic, add an artistic touch to dishes, elevating the dining experience.

Creative Reinvention

Modern chefs don't limit themselves to traditional recipes. They often reinvent classics by incorporating contemporary ingredients, such as figs, to create dishes that honor tradition while capturing the spirit of innovation.

Cultural Resonance

The history of figs in Mediterranean culture adds deep resonance to these modern dishes. Figs, closely linked to the cultural identity of the region, add a cultural and emotional dimension to meals.

Figs, symbols of the Mediterranean, blend harmoniously into the region's modern cuisine.

By providing natural sweetness and rich flavors, they broaden the culinary palette while celebrating the gastronomic traditions of the past. In modern Mediterranean cuisine, figs embody the continuity of culinary history while injecting a touch of delectable modernity.

Chapter 85: Sweet Wisdom: The Fig Tree in Ancient Spiritual Teachings

For millennia, ancient spiritual teachings have found profound symbols and metaphors in nature to convey timeless lessons. Among the natural elements that have captured the attention of thinkers and spiritual teachers, the fig tree stands out for its rich symbolism and its role in expressing spiritual teachings. In this chapter, we will explore how the fig tree has been integrated into ancient spiritual teachings, offering lessons of growth, wisdom, and spiritual connection.

The Fig Tree as a Metaphor of Spiritual Growth

The fig tree, with its gradual growth process, has been used as a metaphor for spiritual evolution. Ancient teachings often compare the growth of figs to the maturation of the human soul, emphasizing that spiritual understanding develops slowly and flourishes over time.

The Leaves of the Fig Tree: Symbol of Protection and Knowledge

In some traditions, the leaves of the fig tree are considered a symbol of protection and knowledge. The biblical story of Adam and Eve, who covered themselves with fig leaves after realizing their nudity, is interpreted as a quest for wisdom and spiritual discernment.

The Similarity between the Fig Tree and the Human Being

Ancient spiritual teachings have often emphasized the similarity between the fig tree and the human being. As the fig tree produces sweet fruit, individuals are encouraged to cultivate gentle inner qualities, such as kindness, compassion and love.

The Resilience of the Fig Tree in the Face of Adversity

The fig tree, known for its resilience in difficult environments, has inspired teachings about perseverance in the face of spiritual challenges. Spiritual teachers have used the fig tree to remind disciples of the importance of remaining strong and committed despite obstacles.

The Spiritual Harvest Metaphor

The harvest of figs, season after season, is often used as a metaphor for the spiritual harvest. Ancient teachings compare the harvest of fruit to the harvest of virtuous qualities and spiritual knowledge accumulated over time.

The Fig Tree as a Symbol of Cosmic Connection

In some spiritual traditions, the fig tree is considered a symbol of the connection between the cosmos and the individual soul. The branches of the fig tree, which extend widely, are interpreted as a reminder that the soul is interconnected with the universe.

The Importance of Now

Figs, ripening quickly, often symbolize the importance of the present moment. Ancient spiritual teachings remind us that wisdom and understanding come from fully immersing ourselves in the current moment, just as tasting a sweet fig is an experience to be savored fully.

The fig tree, with its complex symbolism and natural attributes, has found its place in ancient spiritual teachings as a source of inspiration and reflection. Through visual metaphors and practical lessons, the fig tree has been a powerful means of conveying deep spiritual truths. Connecting to the growth, resilience and gentleness of the fig tree, ancient spiritual teachings invite us to meditate on our own spiritual path and find wisdom in the beauty of nature around us.

Chapter 86: Guardians of the Earth: Fig Trees and Soil Conservation

In the complex fabric of the Earth's ecosystem, soils play a vital role in providing nutrient support for vegetation and harboring essential biodiversity. Among the actors that contribute to the preservation of these precious soils, fig trees stand out for their symbiotic interactions and their abilities to maintain soil health. Let's examine how fig trees become true guardians of the earth by preserving the soil and supporting ecological balance.

Fig Trees and Soil Erosion

One of the major contributions of fig trees to soil conservation is their ability to reduce erosion. The root systems of fig trees are known for their depth and breadth, which adds soil stability and prevents erosion caused by strong winds and torrential rains.

The Formation of a Beneficial Microclimate

Fig trees have the ability to create a favorable microclimate around them. Their leaves provide shade and slow evaporation, creating wetter conditions in the soil. This helps maintain soil moisture, which is essential for its fertility and ability to support plant growth.

Soil enrichment

Fig trees have the ability to fix atmospheric nitrogen in the soil through symbiotic associations with nitrogen-fixing bacteria. This nitrogen fixation enriches the soil with essential nutrients, promoting the healthy growth of neighboring plants.

Creation of a Conducive Habitat

Fig trees, with their complex root systems and lush foliage, create environments conducive to underground life. Microorganisms, insects and small animals find refuge in the fertile soil beneath the fig trees, contributing to the overall health of the ecosystem.

Improvement of Soil Quality

Falling leaves from fig trees and rotting fruit contribute to the formation of rich organic litter. This organic matter decomposes over time, improving the soil's structure, ability to retain water and its nutritional properties.

Beneficial Ecological Interactions

Fig trees often establish symbiotic relationships with other plants and trees. These associations promote biodiversity by creating varied habitats and stimulating ecological interactions that benefit the entire ecosystem.

The Inspiration of Harmonious Coexistence

Fig trees, with their ability to improve soils and promote ecological balance, offer inspiration for harmonious coexistence between man and nature. They remind us that soil preservation is a collective and essential task to ensure the health of the planet.

Fig trees, with their significant contributions to soil preservation, embody the vital role that trees play in maintaining ecological balance. By reducing erosion, enriching the soil and creating beneficial microclimates, fig trees become guardians of the earth, contributing to the sustainability of ecosystems and the health of our environment. They remind us that every living being has a crucial role to play in preserving the earth that shelters us and in protecting the soil that supports life.

Chapter 87: Tasty Heritage: The Fig in the Culinary Traditions of Latin America

Latin America is a vibrant mosaic of cultures, traditions and flavors. At the heart of its rich and diverse cuisines, the fig presents itself as an iconic ingredient that connects the past and the present through its varied culinary uses. Let's dive into the culinary traditions of Latin America and

We will discover how the fig has been inserted with grace and delight into the traditional and contemporary recipes of the region.

Historical Roots

Fig trees were already present in Latin America long before the arrival of Europeans, leading to the seamless integration of this fruit into indigenous cuisines. Pre-Columbian cultures found figs a valuable source of nutrition and explored innovative ways to incorporate them into their diets.

Main Course and Side dish

In some regions, figs have been used as the main ingredient in savory dishes. For example, figs stuffed with minced meat, vegetables and spices are a delicacy in some Latin American cuisines, offering a unique balance between sweetness and savory flavor.

Sauces and Marinades

Figs have also found their way into creating sauces and marinades. Their sweet and tangy flavor brings complexity to these preparations, transforming meat and fish dishes into sensational taste experiences.

Traditional Pastries

In many Latin American countries, figs are made into jams, fruit jellies and traditional desserts. These sweets are often prepared on special occasions and celebrate the abundance of the harvest.

European Influence

Spanish and Portuguese influence in Latin America also introduced the use of figs in pastries and desserts. Dried figs are frequently used to add natural sweetness cakes, cookies and pies.

Modernity and Innovation

The fig continues to play a central role in modern Latin American cuisines. Contemporary chefs are exploring new flavor combinations by integrating figs into fusion and creative dishes that combine ancestral traditions and contemporary influences.

A Symbol of Conviviality

In Latin America, cooking is much more than a simple necessity, it is an expression of conviviality and sharing. Figs, with their gourmet nature and delicious taste, are often associated with these moments of sharing around the table.

The Evolution of Cuisine

The presence of figs in the culinary traditions of Latin America testifies to the evolution of cuisine over the centuries. This versatile fruit has witnessed culinary changes while preserving the roots and richness of traditional flavors.

In the culinary traditions of Latin America, the fig shines as a gastronomic jewel, bearer of history and exquisite tastes. Its role in savory dishes, pastries, jams and desserts is a testament to its versatility and adaptability. The fig continues to link the past with the present, connecting cultures and generations around the shared love of cuisine and conviviality.

Chapter 88: Holistic Balance: Fig in Ayurvedic Medicine Practices

Ayurveda, an ancient health science originating from India, considers nutrition a fundamental pillar of well-being. Among the many ingredients that enrich Ayurvedic practices, the fig stands out for its medicinal properties and its contributions to the holistic balance of body and mind. Let us examine how the fig found its valuable place in the arsenal of Ayurvedic medicine, providing a unique perspective on health and wellness.

Role in the Doshas

Ayurveda identifies three doshas, or biological forces, that influence a person's health and temperament: Vata, Pitta and Kapha. Fig, with its sweet and refreshing nature, is often recommended for balancing the Pitta and Vata doshas by calming the inner fire and promoting relaxation.

Digestion and Absorption

Figs are considered foods that support digestion. Their combination of soluble and insoluble fiber promotes regular intestinal transit, helping to eliminate toxins and maintain digestive balance.

Antioxidant Properties

Figs are rich in antioxidants, such as polyphenols and flavonoids, which neutralize free radicals in the body. These antioxidant properties help strengthen the immune system and prevent cell damage.

Strengthening the Immune System

Fig is a source of essential vitamins and minerals, including vitamin C, potassium and calcium, which strengthen the immune system and support bone health.

Fluid Balance

Fresh and dried figs are rich in fiber and potassium, which helps maintain body fluid balance. This fluid balancing helps prevent swelling and maintain adequate hydration.

Use in Herbal Teas and Infusions

Figs can be used to prepare herbal teas and infusions with soothing properties. They are often combined with other Ayurvedic ingredients such as spices to create drinks that nourish and comfort.

Harmonization of the Spirit

Ayurveda recognizes the interconnection between body, mind and soul. Figs, by providing nutritional and energetic support, play a role in harmonizing these aspects of being, promoting a feeling of overall well-being.

Respect for the Season and the Individual

A key feature of Ayurveda is respect for the seasons and individual needs. Figs, being seasonal and adapted to warm temperatures, are integrated into Ayurvedic diets depending on climatic conditions and individual predispositions.

The fig, rooted in the fertile soil of Ayurvedic medicine, embodies the holistic principles of this ancient tradition. As an ingredient that supports digestion, strengthens the immune system and balances the doshas, fig speaks to the depth of Ayurvedic wisdom. Its integration into diets and wellness practices reflects Ayurveda's comprehensive approach to health, aiming to balance the body, mind and soul for vitality. and lasting harmony.

Chapter 89: Between the Worlds: Fig Trees in Nordic and Celtic Myths

The myths and legends of Nordic and Celtic peoples are woven with deep connections between nature and the divine. Among the natural elements that play a significant role in these stories, fig trees stand as trees of symbolism and mystery. In this chapter, we'll delve into Norse and Celtic myths to discover how fig trees were incorporated into these epic tales and how they embody concepts such as spirituality, protection, and passage between worlds.

The World Tree Yggdrasil in Norse Mythology

In Norse mythology, the world tree Yggdrasil is a monumental oak tree that connects the nine worlds of the universe. Although the fig tree does not appear directly in this mythology, its symbolism of the sacred tree and the link between worlds is echoed in the representation of Yggdrasil, emphasizing the importance of trees in cosmology Nordic.

The Fig Tree of Fal in Celtic Mythology

Celtic mythology is rich in stories where fig trees play a prominent role. The "Fal Fig Tree" is a notable example. According to legend, the fig tree is found in Tara, a sacred place in Ireland. When a claimant to the throne stood on a stone called Lia Fáil, the fig tree would have grown or flowered to confirm his legitimacy as king. This interaction between the fig tree and the place of power underlines the role of the tree as divine witness and judge.

The Fig Tree as a Portal Between the Worlds

In Celtic myths, fig trees are sometimes considered portals between the world of the living and that of spirits. They are associated with the "sídhes", mystical hills where fairies and spirits reside. These sacred fig trees act as points of contact between earthly and spiritual realities, symbolizing the connection between the worlds.

The Symbolism of Protection

Fig trees, with their deep roots and imposing stature, are often seen as symbols of protection in myths. Fig trees offer their shade, shelter and energy to those seeking refuge, reinforcing the idea of the tree as guardian of souls.

The Alliance between Man and Nature

Norse and Celtic myths highlight the sacred alliance between man and nature. Fig trees, as sacred natural elements, embody this relationship, serving as a reminder that human beings are deeply connected to the natural world and its mysteries.

The Role of the Fig Tree in Heroic Stories

Fig trees also appear in the heroic stories of these cultures, often as magical or symbolic elements. They can represent challenges to overcome, divine advice or reference points in the hero's quest.

Fig trees, loaded with symbolism and mystical power, fit gracefully into Norse and Celtic myths. As gatekeepers, protectors and witnesses to extraordinary events, they embody the intimate connection between man and nature in these ancient traditions. Fig trees are living reminders of the importance of respecting and preserving the balance between the physical and spiritual worlds, in which the tree becomes a guide between the hidden mysteries of the universe.

Chapter 90: Brilliant Gourmandise: The Fig in Vegetarian Gastronomy

Vegetarian gastronomy is a celebration of the richness of natural flavors, where vegetables, fruits and plants form the taste palette. Among vegetarian culinary gems, the fig shines as a versatile and delicious ingredient. The fig has won the hearts of vegetarian food lovers by bringing a touch of elegance and flavor to a variety of dishes.

A Visual and Taste Feast

The fig, with its velvety skin and fleshy flesh, brings a visual dimension to vegetarian gastronomy. Its attractive appearance creates a visual effect that catches the eye and stimulates the appetite, enhancing the dining experience.

Amalgamation of Flavors

Fig offers a unique combination of natural sweetness and slightly tart notes. This juxtaposition of flavors allows figs to pair harmoniously with a variety of ingredients, from cheeses to nuts to green vegetables.

In Savory Dishes

Figs add a sweet, decadent touch to savory dishes. They can be roasted to concentrate their flavors or served chilled for a refreshing contrast. Figs go perfectly with salads, vegetarian pizzas and grain dishes.

Celebration of Vegan Cheeses

Figs and vegan cheeses are a dream duo. Figs provide a natural sweetness that balances the richness of vegan cheeses, creating a symphony of textures and flavors in every bite.

In Pastas and Risottos

Figs become star ingredients in vegetarian pastas and risottos. They add a sweet and delicate dimension that complements the richness of sauces and rice preparations.

Radiance in Desserts

Figs are undisputed stars in vegetarian desserts. They can be used to create exquisite tarts, cakes, jams and compotes that delight the taste buds and bring a sweet touch to the end of a meal.

Incorporation into Beverages

Figs can also be included in vegetarian drinks. Fig-based smoothies, juices and teas provide a unique natural sweetness and depth of flavor.

A Source of Nutrition

In addition to their divine taste, figs are also rich in nutrients. They are an excellent source of fiber, vitamins and essential minerals, contributing to a balanced vegetarian diet.

The fig, with its captivating taste and its ability to transform vegetarian dishes into exquisite delights, finds a special place in vegetarian gastronomy. It adds a touch of elegance and originality to recipes while providing essential nutrients. Figs, true jewels of nature, celebrate culinary creativity and enrich vegetarian meals with an incomparable taste experience.

Chapter 91: The Ecological Renaissance: Fig Trees and Ecosystem Restoration

In the ever-changing world, ecosystem restoration has become a crucial priority to maintain environmental balance. Fig trees, with their unique properties and ecological role, are emerging as key players in the preservation and restoration of ecosystems. Let us ask ourselves how fig trees have been involved in ecological restoration, contributing to the regeneration of degraded lands and the reconstitution of biodiversity.

Pioneers of Restoration

Fig trees are often called "pioneer trees" due to their ability to quickly colonize degraded soils. Their deep roots help prevent erosion and stabilize soils, creating conditions for other plants to regrow and revitalize ecosystems.

Mycorrhiza Partners

Fig trees establish symbiotic relationships with mycorrhizal fungi. These fungi help improve soil structure, facilitate the flow of nutrients, and promote the growth of surrounding plants. Thus, fig trees act as "ecosystem engineers" by creating an environment favorable to the restoration of biodiversity.

Wildlife Attractors

Fig trees also play a vital role in attracting wildlife. Their sweet fruits provide a food source for a variety of animals, such as birds, bats and small mammals. By attracting these creatures, fig trees participate in seed dispersal and renewal of ecosystems.

Protection Against Desertification

In regions prone to desertification, fig trees can play a crucial role in preventing this threat. Their extensive root systems help maintain soil moisture and prevent the spread of dry land, helping to preserve areas affected by degradation.

Rehabilitation of Urban Areas

Fig trees are also used in the rehabilitation of degraded urban areas. Their ability to thrive in harsh environments makes them ideal candidates for revegetating urban spaces, improving air quality, providing shade and creating habitats for wildlife.

Restoring Biodiversity

Fig trees act as "beacons of biodiversity", attracting a multitude of plant and animal species. By providing resources and habitats, fig trees help restore ecological balance and promote the harmonious coexistence of living beings.

Fig trees, with their multifunctional role in ecosystem restoration, are true allies in the quest for environmental preservation. Their ability to revitalize soils, attract wildlife and create ecological niches make them crucial elements in the reconstitution of fragile ecosystems. Fig trees eloquently illustrate nature's potential to self-regenerate, offering a glimmer of hope in global efforts to restore and preserve the beauty and diversity of our planet.

Chapter 92: The Carved Essence: The Art of Fig Wood Carving

The art of wood carving has a long history, transcending cultures and eras to bring to life works of timeless beauty. Among the precious woods used for this form of artistic expression, the fig tree stands out for its distinctive texture and unique character. In this chapter, we

We'll delve into the art of fig wood carving, exploring how this material provides a unique canvas for creativity and artistic expression.

The Elegance of Material

Fig wood, with its swirling patterns, organic veins and contrasting areas of light and dark, provides an inspiring visual palette for carvers. Each piece of fig wood is a work of art in itself, bearing the traces of the tree's growth and time.

The Dance of Nature

The natural patterns of fig wood evoke the organic, shifting aspect of nature itself. Fig wood carvings often capture biomorphic forms, reflecting the way life takes fluid and changing forms in nature.

Meticulous Artistic Work

Fig wood carving requires a blend of technical expertise and artistic creativity. Carvers must take into account variations in grain, varying hardness, and specific properties of wood to create works that transcend the limitations of the material.

An Intimate Connection with Material

Fig wood carvers often develop an intimate relationship with the material. They listen to the stories that the wood tells them through its patterns, adapt to the whims of the grain and give birth to creations that are a tribute to the essence of the tree.

A Dialogue between Sculptor and Wood

Fig wood carving is a dialogue between the artist and the material. The sculptor works in harmony with the wood, finding shapes that blend with its natural characteristics while creating something new and beautiful.

The Valorization of Imperfection

Fig wood carvings often celebrate the imperfections and irregularities of the material. Knots, cracks and unusual patterns become unique design elements, adding character and depth to the finished work.

A Cultural and Artistic Heritage

Fig wood carving is often rooted in cultural and artistic traditions. In some cultures, the fig tree is considered sacred, and fig wood carvings can carry deep cultural meanings, conveying stories, beliefs and values.

The art of fig wood carving transcends the boundaries between art and nature, artist and material. Each work results from a collaboration between the sculptor and the wood, capturing the very essence of the tree and human creativity. Fig wood carvings are testaments to natural beauty, artistic mastery, and the deep connection between man and nature, creating works that will continue to inspire and amaze people. future generations.

Chapter 93: Between Worlds: The Fig in Contemporary Fantasy Literature

Contemporary fantasy literature explores the unexplored recesses of the human imagination, weaving tales that challenge the limits of reality. Among the elements evoked in these extraordinary worlds, the fig emerges as a symbol rich in mystery and symbolism. Let's examine how the fig finds its place in contemporary fantasy literature, as an element that transcends the boundaries of reality and opens doors to enchanting universes.

Magical Portal to the Unknown

In fantasy literature, the fig is often used as a mystical portal to other worlds. The protagonists can enter a parallel universe by crossing a fig tree, thus creating a link between the tangible world and the fantasy realm. The fig tree becomes a symbol of the bridge between reality and the extraordinary.

The Enchanted Fig

In some fantastic stories, the fig is presented as an enchanted fruit, endowed with magical powers. Characters can be transformed, healed, or given exceptional knowledge by consuming special figs. This depiction highlights the mystical nature of the fruit and its potential to change the course of destiny.

The Magic Garden of Fig Trees

Fig gardens in fantasy literature often become havens of magic and secrets. These gardens are places where time bends, where extraordinary creatures reside, and where reality is shaped by the desires and dreams of the characters. Fig trees, with their unique characteristics, become the guardians of these enchanted gardens.

The Protective Fig Tree

In some fantasy stories, fig trees are presented as guardians and protectors of hidden secrets. Their intertwined branches and dense shadow create a safe sanctuary for characters, helping them escape dark forces or find refuge in uncertain worlds.

The Symbolism of Transformation

Fig trees in fantasy literature can symbolize the transformation and evolution of characters. As the tree itself goes through cycles of growth and change, characters can find mirrors of their own journey through the fig trees, inspiring their own quest for personal discovery.

The Duality of Reality

The fig in fantasy literature often embodies the duality of reality and imagination. THE

Characters can get lost among the leaves of a fig tree, travel between worlds and question what is real and what is not, inviting readers to explore the boundaries of perception.

The fig in contemporary fantasy literature is much more than just a fruit. She becomes a symbol of the unknown, magic and transformation, adding depth and complexity to fantasy worlds. Fig trees act as portals, guardians and catalysts of the marvelous, inviting readers to cross the boundaries of reality to explore the fascinating recesses of the imagination.

Chapter 94: Urban Development: Fig Trees in Urban Gardens

In the heart of bustling cities, where concrete and steel dominate the landscape, urban gardens arise like green oases, providing a refuge for nature in the midst of frenetic urbanity. Among the elements that make their way into these green spaces, fig trees emerge as symbols of connection with nature and a link between past and present. Fig trees fit harmoniously into urban gardens, bringing a touch of rusticity and serenity to the city environment.

The Return to Nature in an Urban Environment

Fig trees, with their lush leaves and organic appearance, embody the return to nature in the heart of the urban jungle. Their presence in urban gardens offers residents an opportunity to temporarily disconnect from the hectic pace of urban life and reconnect with the tranquility that only nature can provide.

Creators of Ecological Balance

Fig trees in urban gardens are not only decorative elements, but they also play an essential role in ecological balance. Their leaves provide shade, helping to reduce the urban heat island effect, while their roots help prevent soil erosion and maintain its quality.

Link with the Historical Past

Fig trees have a long history and deep cultural significance, particularly in Mediterranean regions. By integrating fig trees into urban gardens, designers and planners are forging a subtle connection to the past, recalling ancient traditions while building a sustainable future.

Celebration of Biodiversity

Fig trees, by attracting a variety of birds, insects and other creatures, contribute to the biodiversity of urban gardens. They provide habitats for wildlife and create a miniature ecosystem, reminding city residents of the richness of natural life.

Food for the Soul and Body

The presence of fig trees in urban gardens can also have tangible benefits for residents. Ripe figs are a delicious reward, inviting passersby to pick a fresh fruit from the garden and enjoy its nutritional benefits.

Mediation between Opposites

Fig trees, with their enigmatic beauty, create harmony between the contrasting elements of nature and the city. They connect the verticality of skyscrapers to the horizontality of the earth, forming a visual bridge between urban artifice and natural reality.

Fig trees in urban gardens are much more than decorative trees. They act as ambassadors of nature, offering a space of peace and reflection amidst the urban tumult. Their presence reflects our innate desire for connection with the natural world, while providing ecological and aesthetic benefits to urban spaces. By incorporating fig trees into urban gardens, we weave a living connection between the past, present and future, creating green havens that enrich our lives and make cities more sustainable and balanced.

Chapter 95: The Gentleness of Nature: Fig and Natural Skin Care

Over the centuries, humans have sought to draw on nature for beauty and well-being solutions. Among the natural treasures that have captured the attention of skincare enthusiasts, fig emerges as a valuable ingredient, offering an abundance of benefits for the skin. Fig has become a key component of natural skincare, providing caring softness for our outermost shell.

A Source of Antioxidants

Figs are full of antioxidants, these powerful molecules that fight free radicals responsible for premature aging of the skin. Fig extracts in skin care products can help reduce signs of aging and maintain glowing, youthful skin.

Deep Hydration

Fig is naturally full of water, making it a naturally effective moisturizer for the skin. Fig skincare products help to deeply hydrate, soothe dry skin and prevent moisture loss, leaving skin soft and supple.

Gentle Exfoliation

The natural enzymes found in figs can gently exfoliate the skin, removing dead cells and revealing a brighter complexion. Fig-based exfoliating products can help refine skin texture, reduce blemishes and promote cell turnover.

Treatment of Skin Conditions

Fig is also known for its anti-inflammatory and soothing properties. It can be used to calm irritation, relieve redness and soothe sensitive skin. Fig products can help treat skin conditions such as eczema and dermatitis.

Natural Brightness

Fig is rich in vitamins and minerals essential for skin health. Nutrients like vitamin C help brighten skin tone, reduce dark spots, and give skin a natural, radiant glow.

A Fusion of Nature and Beauty

The use of fig in skincare is part of the growing trend among consumers to seek more natural and environmentally friendly products. Fig products offer a pleasant sensory experience while establishing a connection with nature and its benefits.

Ethical Sustainability

Figs, as natural ingredients, also contribute to sustainability and environmental responsibility. Skincare companies that include fig extracts in their products highlight eco-friendly practices and encourage responsible consumption.

The fig, this fruit full of sweetness and benefits, has found its precious place in the world of natural skin care. By incorporating fig into our beauty rituals, we connect with the power of nature to nourish, soothe and beautify our skin. Fig skincare offers a holistic experience, a harmony between the ancient knowledge of natural benefits and the contemporary demands of sustainable beauty.

Chapter 96: Exotic Delights: Figs in Asian Culinary Traditions

Asian culinary traditions are a true taste journey, offering a rich variety of unique flavors, ingredients and cooking techniques. Among the region's gastronomic treasures, figs have found their place as a versatile and delectable ingredient. Figs fit seamlessly into Asian culinary traditions, adding a sweet, sumptuous touch to already famous dishes for their complexity and diversity.

A Fusion of Tastes and Cultures

Figs, although traditionally associated with regions
Mediterranean, have made their way into Asian cuisines to create unexpected and delicious flavor
combinations. They embody a fusion between cultures, connecting distant lands through gastronomic
pleasure.

The Subtle Balance of Flavors

In Asian culinary traditions, the balance of flavors is essential. Figs, with their natural sweetness, bring a
subtle sweet note to dishes that contrasts with the salty, spicy and sour flavors characteristic of Asian
cuisine.

The Presence in Savory and Sweet Dishes

Figs are versatile, being able to be used in a variety of savory and sweet dishes. They can be incorporated
into meat dishes, stews, salads, rice dishes and even soups. Plus, they add a touch of sweetness to traditional
desserts such as pastries, jellies and jams.

Harmony with Spices

Figs combine harmoniously with spices and aromatic herbs often used in Asian cuisine. They can
balance the heat of chili peppers, enhance the flavor of curries and add a touch of elegance to spicy
dishes.

Artistic Presentation

Presentation is a crucial element in Asian cuisine, and figs provide an appealing aesthetic to dishes. Their bright
colors and distinct shape add visual flair to Asian culinary art,

creating a striking contrast with other ingredients.

Celebration of the Seasons

In some Asian cuisines, figs are celebrated based on their seasonality. They are used when they are freshest and most abundant, adding a seasonal dimension to dishes and festivities.

A New Culinary Paradigm

The integration of figs into Asian culinary traditions is a testament to the creativity of chefs and cooks who push the boundaries of tradition while respecting cultural roots. Figs add a new dimension to classic dishes and contribute to the evolution of modern Asian cuisine.

Figs, with their natural sweetness and versatility, have become a harmonious part of Asian culinary traditions. By incorporating this unique ingredient, chefs and cooks create a symphony of taste that celebrates the variety and diversity of Asia's flavors. Figs continue to enrich the region's culinary repertoire, adding an exquisite sweet note to dishes already steeped in history, culture and innovation.

Chapter 97: The Resurrection of the Earth: Fig Trees and the Regeneration of Dry Lands

Drylands, degraded by harsh climatic conditions and inappropriate use, are often considered desolate and barren areas. However, among the many miracles of nature, fig trees emerge as agents of regeneration capable of transforming these distressed landscapes into havens of life. Fig trees play a crucial role in the regeneration of drylands, revealing their incredible ability to breathe life where it seemed lost.

A Botanical Miracle

Fig trees are pioneers in arid regions, blessed with the ability to establish themselves in poor soils and withstand extreme environmental conditions. Their deep roots and ability to store water make them ideal candidates for restoring ecological balance in degraded areas.

Moisture and Nutrient Suppliers

Fig trees, through their transpiration process, release moisture into the surrounding atmosphere, creating a more humid microclimate around them. This increased humidity can encourage the growth of other plants, helping to restore ecosystems.

Additionally, fig trees produce nutrient-rich leaves and fruits, which fall to the ground and decompose, enriching the soil with organic matter and essential minerals.

Hosts for Biodiversity

Fig trees also play a crucial role in providing habitat and food for a variety of wildlife. Birds, insects and small mammals are attracted to fig trees to feed on their fruits, leaves and associated insects, helping to restore the local food chain.

Fight against Erosion and Desertification

In arid regions, erosion and desertification are major problems. Fig trees, with their extensive root systems, can stabilize soils and prevent erosion. Their roots help retain moisture and protect soils from strong winds and intense rainfall.

Restoration of Ecological Balance

When fig trees become established in dry lands, they create a beneficial cascading effect. Their actions encourage the growth of other plants, attracting more animals and thus creating an ecosystem that naturally regulates ecological cycles.

Fig trees stand as ambassadors of hope in dry lands. Their ability to regenerate soils, create more favorable microclimates and serve as pillars for biodiversity illustrates their vital role in restoring distressed ecosystems. By collaborating with fig trees, we can learn from nature itself how to revitalize lands that seem desolate, reminding us that life has the power to flourish even in the most inhospitable conditions.

Chapter 98: Ancient Knowledge and Wisdom of Fig Trees: The Fig in the Tales of Oriental Wisdom

Eastern wisdom tales have been woven through the centuries to convey profound lessons about life, spirituality and human nature. Among the evocative symbols that dot these stories, the fig tree emerges as a recurring element, carrying a rich and complex meaning. In oriental wisdom tales the fig occupies a central place, revealing the universal truths it embodies.

The Fig Tree as a Metaphor of Knowledge

In many oriental tales, the fig tree is represented as a tree of knowledge and enlightenment. Its large, abundant leaves symbolize the vast expanse of wisdom and understanding that can be gained in the quest for truth.

The Search for Inner Wisdom

In these tales, the fig tree becomes a refuge for wise men and spiritual seekers who isolate themselves under its shade to meditate and seek inner truth. The fig tree thus represents a place of retreat, where one can find the tranquility necessary to delve deeply into existential questions.

The Cycle of Life and Death

Fig trees, with their cycles of growth, fruiting and rest, reflect the natural cycles of life and death. In oriental tales, the fig tree is often used to remind readers of the transient nature of human existence and the importance of embracing each moment.

The Fig as a Symbol of Generosity

Eastern wisdom tales often show fig trees offering their fruit to all who seek them. This generosity symbolizes the importance of sharing knowledge, wisdom and blessings with others, evoking the idea of spiritual abundance.

The Fig and the Quest for Truth

In some stories, the fig is used to illustrate the never-ending quest for truth. Figs, with their sweet, juicy interiors, hide tiny seeds, symbolizing the search for the depth hidden behind superficial appearances.

The Fig as a Balance Reminder

Fig trees, with their connection to the earth and their search for the sun, illustrate the importance of balance between the spiritual and the material. The tales highlight how fig trees flourish when they receive care from both heaven and earth.

À Through tales of oriental wisdom, the fig tree becomes much more than just a tree. He is a symbol living from the quest for knowledge, generosity, truth and balance. These timeless stories remind us that wisdom can be found in nature, and that the fig tree, with its beauty and mysteries, guides us to universal truths that transcend the boundaries of time and culture.

Chapter 99: Fig Trees as Pillars of Sustainability: Agroforestry in the Service of the Ecosystem

Agroforestry, an integrated approach to land use, links agriculture with forestry to create productive and sustainable ecosystems. Among the key players in this strategy, fig trees stand out for their ability to promote soil regeneration, promote biodiversity and support the livelihoods of local communities. In this chapter, we will explore the vital role of fig trees in sustainable agroforestry, highlighting their potential to create a harmonious balance between food production and environmental preservation.

Establishing Beneficial Coexistence

Agroforestry is based on the idea of the interdependence between trees and crops. Fig trees, with their ability to grow in marginal soils and tolerate harsh conditions, provide a strong structure for agroforestry systems. They create favorable microclimates for crops by regulating temperature, humidity and light.

Soil Restoration and Erosion Prevention

Agroforestry systems incorporating fig trees contribute to the regeneration of degraded soils. The deep roots of fig trees help retain moisture and prevent erosion, creating conditions suitable for crop growth and long-term preservation of farmland.

Biodiversity and Wildlife Habitats

Fig trees act as focal points for biodiversity. Their fruits, leaves and branches attract a variety of animals, such as birds, insects and small mammals. This biological diversity contributes to the balance of the ecosystem, promoting pollination, the regulation of pests and the renewal of nutrients.

Livelihood Support

Fig trees in agroforestry systems can provide a source of income and food for local communities. The fruits can be sold in markets, transformed into by-products or used for family consumption. Additionally, fig trees provide shade for livestock and thus contribute to animal husbandry.

Environmental Education

Agroforestry with fig trees also provides opportunities for environmental education. Local communities can learn about sustainable practices, the importance of preserving the biodiversity and the interconnection of natural elements in their environment.

Fig trees stand at the heart of sustainable agroforestry, illustrating how a balanced approach between agriculture and forestry can create thriving ecosystems. By promoting soil restoration, biodiversity, land regeneration and supporting local communities, fig trees embody the very essence of agroforestry. Their presence nourishes the soil, supports wildlife and strengthens the relationships between humans and their environment.

Chapter 100: Elegance in Miniature: Fig Trees in the Art of Bonsai

The art of bonsai, an ancient art form originating from Asia, embodies the beauty and harmony of nature through the cultivation of small trees in pots. Among the species prized for this delicate practice, fig trees stand out for their adaptability, attractive foliage and potential to evoke the grandeur of nature in a small space. fig trees have taken the bonsai world by storm, offering a captivating insight into the symbiosis between man and nature through the creation of these revered miniatures.

The Expression of Life in Miniature

The fig bonsai is much more than just a potted plant. It is a work of art that encapsulates the spirit of a mature tree in a small space. The fig tree, with its distinct characteristics and twisted trunk, provides a perfect canvas for bonsai artists to express the beauty and vitality of nature.

Time and Patience

Creating a fig bonsai requires extreme patience. By growing a young tree to resemble its older counterpart in nature, bonsai artisans create a work that tells the story of time and growth.

A Balance between Precision and Naturalness

The art of bonsai relies on the balance between technical precision and natural appearance. The fig trees, with their tortuous shapes and dense foliage present unique challenges. Bonsai artists must shape trees carefully, respecting their natural growth while giving them an elegant aesthetic.

Spiritual Symbolism

In many Asian cultures, fig trees symbolize longevity, prosperity and wisdom. Bonsai fig trees embody these attributes, providing a constant reminder of the importance of connecting with nature and respecting its cycles.

Learning Humility

Growing bonsai fig trees teaches humility. Artists learn to work in harmony with the rhythms of nature, to listen to the trees and to adapt to the specific needs of each specimen. This learning process reminds us that even in art, man is ultimately collaborating with natural greatness.

Fig trees have earned their place in the exquisite art of bonsai as living symbols of beauty, patience and harmonious coexistence with nature. Bonsai artists enjoy transforming these trees into miniatures that convey a message of respect for nature and celebration of life. Bonsai fig trees will continue to captivate and inspire future generations with their timeless elegance and ability to connect humans to the splendor of nature in miniature.

Chapter 101: A Feast of Flavors: Fig in Indian Culinary Traditions

India, rich in cultural and culinary diversity, has always been a melting pot of flavors and traditions. At the heart of this varied taste palette, the fig stands out as a precious ingredient, both for its delicious taste and for the symbolic meanings it embodies. In this chapter, we delve into Indian culinary traditions where the fig takes pride of place, revealing how it weaves a thread between food, culture and the hearts of Indians.

Fig: A Treasure of Natural Sweetness

Figs, with their sweet, juicy flesh, add a natural sweetness to Indian dishes. They are used in a variety of preparations, from spicy curries to decadent desserts, adding a subtle flavor and a pleasant touch of sweetness.

Dried Figs in Indian Cuisine

Dried figs, also called anjeer, have a pride of place in Indian desserts. They are often used to prepare barfis (sweets made from milk and sugar), halwas (sweet semolina cakes) and laddus (sweet spheres). Dried figs lend a chewy texture and natural sweetness to these creations.

The Symbolism of the Fig

In India, the fig is associated with prosperity, health and fertility. It is often offered as an offering in temples and used during religious and family celebrations. This symbolism strengthens the link between the fig and Indian culture, making this fruit a meaningful ingredient in culinary and spiritual practices.

Creative Uses in the Kitchen

Figs are also used in savory dishes like chutneys and condiments. They add a sweet-spicy note to traditional accompaniments, creating a balance of flavors that is characteristic of Indian cuisine.

The Influence of Regions

India's regional diversity is reflected in the use of figs. In Kashmir, for example, they are a key ingredient in the preparation of rich and aromatic dishes. In South India, they can be used in coconut and rice dishes.

The Fig as Cultural Metaphor

The fig, with its variety of preparations and meanings, becomes a cultural metaphor for India itself: diverse, complex and imbued with deep cultural richness.

Indian culinary traditions are a celebration of the diversity of flavors and symbols. The fig, with its delectable sweetness and powerful symbolism, weaves a thread between cuisine, culture and spirituality. Its creative use in both sweet and savory dishes embodies the richness of Indian cuisine, while reminding us that food can be much more than just a taste experience – it can be an expression of culture and the heart.

Chapter 102: The Art of Traditional Cultivation of the Fig Tree: Methods Passed Down From Generation to Generation

Generation

Fig cultivation has a long history dating back to ancient times, and traditional cultivation methods have been preserved and passed down from generation to generation. These methods respect the symbiosis between the tree and the environment, while ensuring an abundant and quality harvest. Traditional methods of fig cultivation have persisted through time, reflecting a harmonious relationship between man and nature.

Selection and Planting of Varieties

Traditional growing methods begin with the careful selection of fig varieties. Varieties adapted to the local climate and soil are chosen to guarantee the success of the crop. Fig trees are usually planted in the fall or spring, when conditions are favorable for them to take root.

Choice of Location

Planting location is crucial in traditional fig tree cultivation. A sunny location with well-drained soil is preferred. Fig trees are often planted near walls or buildings to take advantage of warmth and protection against cold winds.

Care and Pruning

The maintenance of fig trees follows a seasonal rhythm. During the first years, particular attention is paid to regular watering to promote rooting. In winter, fig trees are pruned to remove dead branches and promote healthy growth.

Protection Against Diseases and Pests

Traditional methods also include techniques to protect fig trees from diseases and pests. Practices such as crop rotation, application of natural remedies and planting beneficial companions are implemented to maintain the health of the fig trees.

Use of Natural Fertilizers

Traditional farmers often favor the use of natural fertilizers such as compost and manure to enrich the soil with essential nutrients. This approach respects the ecological balance and promotes the long-term health of fig trees.

Harvest and Consumption

The fig harvest is a crucial moment. Farmers carefully observe the color and texture of the fruits to determine their ripeness. Figs are harvested by hand and consumed fresh or transformed into by-products such as jams, preserves or dried ones.

Transmission of Knowledge

Traditional methods of growing the fig tree are often passed down orally from generation to generation. Elders share their knowledge and experience with younger ones, ensuring the sustainability of these valuable methods.

Traditional methods of fig cultivation are a cultural and ecological heritage that honors the relationship between man and nature. These environmentally friendly and climate-responsive approaches have survived over the centuries, testifying to the effectiveness of the harmony with which man can cultivate the land for his livelihood.

Chapter 103: The Fig in North African Cuisine: A Taste Journey Through Flavors

Traditional

North African cuisine, rich in spices, textures and stories, embodies the cultural diversity of the region. Among the precious ingredients that intertwine to create memorable dishes, the fig stands out for its sweet taste, its versatility and its cultural symbolism.

Figs and Culinary Traditions

In Maghreb countries, figs are integrated into a variety of dishes, from starters to desserts. Their natural sweetness pairs perfectly with the bold flavors and spices characteristic of North African cuisine.

Elegant Starters

Fresh or dried figs, accompanied by cheese, nuts and honey, make refined starters that balance textures and tastes. These combinations create symphonies of flavors that awaken the senses and prepare the palates for the feast to come.

Versatile Ingredients

Figs are used versatile in North African cuisine. They can be incorporated into tagines, these emblematic slow-cooked dishes of the region, where they bring a subtle sweetness that contrasts with the spicy flavors of meats and vegetables.

Seasonal Fruits

Fresh figs are often used in seasonal recipes. When plentiful, they become the centerpiece of many dishes, from salads to cakes, jams and pastries.

Figs and Celebrations

In many North African cultures, figs are associated with festive times and celebrations. They are offered as a sign of hospitality and welcome, symbolizing the abundance and generosity shared among friends and family.

A Cultural Heritage

The use of the fig in North African cuisine dates back centuries, testament to the cultural exchanges that have enriched the region over time. It is not only an ingredient, but also a witness to the history and traditions that have shaped this unique cuisine.

The fig, with its sweetness and versatility, is a culinary treasure that delights palates throughout North African cuisine. It demonstrates the importance of the connection between land, culture and food, while highlighting the creativity and passion that animate Maghreb cuisines. In every bite of fig dish, the flavors of the region come together to create a captivating taste experience that celebrates the very essence of North African cuisine.

Chapter 104: Growing Figs in Harmony with Nature: The Secrets of Organic Cultivation

The organic cultivation of the fig tree embodies an approach that respects the earth and its natural cycles. Avoiding chemicals and promoting ecological balance, this traditional and environmentally friendly method preserves the purity of figs while celebrating the relationship between man and nature.

Selection of Adapted Varieties

The first secret of organic fig tree cultivation lies in the choice of varieties adapted to the climate and ground. Opting for varieties that thrive naturally in the area minimizes the need for disease and pest control.

Soil Care and Natural Fertilization

Fig tree health begins with well-nourished soil. Enriching the soil with organic materials like compost and manure promotes healthy growth. Natural fertilizers improve soil structure, increase water retention and provide essential nutrients.

Balanced Watering

Water is essential for the growth of fig trees, but overwatering can lead to root rot problems. Balanced watering, adapted to the specific needs of each tree, preserves the health of fig trees while preserving the precious water resource.

Natural Protection Against Pests

Organic farming promotes the use of natural methods to control pests. Introducing beneficial companion plants, such as aromatic herbs, can repel harmful insects while maintaining a balanced ecosystem.

Thoughtful Pruning and Pruning

Regular pruning of fig trees promotes optimal tree structure, allowing for better air circulation and maximum exposure to sunlight. Proper pruning also helps prevent fungal diseases by removing diseased parts.

Conservation of Local Resources

Organic fig tree cultivation embraces the idea of conserving local resources. Using traditional soil conservation methods, such as terrace planting or the use of plant barriers, helps preserve the balance of the ecosystem.

Education and Transmission

Organic fig tree cultivation often involves educating local farmers and gardeners about environmentally friendly methods. This transmission of knowledge ensures that these valuable methods are preserved for future generations.

Organic fig tree cultivation is a harmonious dance between man and nature, revealing the timeless secrets of sustainable coexistence. By avoiding chemical inputs and highlighting natural cycles, this approach presents a model of sustainability that honors the earth and its gifts. Organic fig tree cultivation is not just an agricultural method, but a philosophy that recognizes the importance of preserving the land for the well-being of future generations.

Chapter 105: Fig and Permaculture in Urban Environments: A Symphony of Sustainability at the Heart of the
City

Permaculture, a holistic approach to sustainable agricultural design, has found a new way into urban environments, and the fig, with its versatility and ability to integrate harmoniously, plays a central role in this journey. Growing figs in an urban environment using permaculture principles opens the way to fascinating possibilities where nature and the city coexist in symbiosis. Figs and permaculture come together to create a nourishing and resilient ecosystem in the heart of metropolises.

Vertical and Horizontal Integration

Urban fig cultivation implements permaculture concepts such as vertical and horizontal integration. Fig trees, with their height and width growth, can be planted along walls, on balconies, or even in community roof gardens, maximizing the use of space.

Biodiversity and Beneficial Interactions

Permaculture encourages the creation of diverse systems that mimic natural ecosystems. Fig trees, as a key part of these systems, attract a variety of beneficial insects and birds, helping with pollination and pest control.

Resource Management

Permaculture principles emphasize intelligent resource management. Fig trees, known for their drought resistance once established, can be fed with collected rainwater or recycled gray water, helping with urban water conservation.

Creation of Favorable Microclimates

Fig trees, with their broad leaves and dense branches, create microclimates conducive to the growth of other plants. These microclimates provide shade, regulate temperature and promote moisture retention, creating an environment conducive to biodiversity.

Community Engagement and Education

Growing figs in an urban environment according to permaculture principles strengthens community engagement and education on sustainability. Fig gardens can become meeting points, educational spaces and sources of inspiration for urban dwellers.

Abundant Harvest and Short Circuits

Figs generally produce an abundance of fruit, which helps promote short circuits by reducing the distance between harvest and consumption. Fresh figs can be shared with neighbors and surplus transformed into local artisan products.

The fig, with its adaptable nature and lush growth, offers an exciting opportunity to integrate permaculture into urban environments. By creating nourishing and ecologically sustainable spaces in the heart of cities, fig cultivation according to the principles of permaculture transcends the traditional boundaries between nature and the urban. It provides a path to a future where nature, community and sustainability come together to create a symphony of thriving life in the heart of the city.

Chapter 106: Sculpting Natural Elegance: Advanced Pruning Practices for Fig Trees

Pruning fig trees is a skill of both artist and scientist, a subtle dance between form and function. Advanced pruning practices transcend simple maintenance to create trees that combine aesthetics, yield and health.

The Art of Architectural Size

Advanced pruning of fig trees is not just about cutting branches, but about sculpting their architecture. Artistic forms, such as the espalier, the flattened crown or the Greek vase, are used to create visually appealing and functional structures.

Optimization of Light and Air

Advanced pruning techniques aim to optimize the circulation of light and air through the tree. Thinning the interior branches allows light to penetrate to the lower parts, promoting balanced growth and avoiding areas of stagnant humidity.

Pruning Practices Depending on Varieties

Each variety of fig tree has specific pruning needs. Some varieties thrive with severe pruning, while others prefer gentler pruning to encourage natural growth. Understanding these nuances is essential to successful advanced pruning.

Size Depending on the Seasons

Advanced pruning of fig trees is a seasonal practice. Winter pruning, when the tree is dormant, promotes rapid wound healing. Summer pruning can be used to control the vigorous growth of branches.

Health and Disease Prevention

Proper advanced pruning promotes tree health by removing dead, diseased, or malformed branches. This reduces the risk of pest and fungal disease infestation, while promoting vigorous growth.

Balance between Production and Aesthetics

Advanced pruning aims to balance fruit production with the aesthetics of the tree. Strategically removing branches helps prevent overproduction that could deplete the tree and reduce the size of the figs.

Contribution to Art and Science

Advanced pruning of fig trees is both an artistic expression and a scientific practice. Advanced pruners understand the individual needs of each tree while creating shapes that beautify the landscape.

Advanced fig tree pruning is an alchemy of skill, creativity and a deep understanding of each tree's needs. She reminds us that nature can be sculpted with care and respect to create living masterpieces that bring both fruit and beauty. Advanced pruning practices for fig trees are a tribute to the co-creation between man and nature, a harmony that transcends seasons and generations.

Chapter 107: Fig in Modern Medical Cuisine: A Tasty Alliance for Health

Modern medical cuisine highlights the power of food to support health and well-being. Among nature's treasures, the fig stands out not only for its delectable flavor, but also for its nutritional and therapeutic properties. The fig, as a valuable ingredient in medical cuisine modern, fits harmoniously into the quest for vitality and well-being.

Essential Nutrients

Fig is rich in essential nutrients, such as fiber, vitamins and minerals. Fiber promotes healthy digestion, while vitamins and minerals strengthen the immune system and support optimal bodily functions.

Natural Antioxidants

Figs are full of antioxidants, such as polyphenols and flavonoids, which neutralize free radicals responsible for premature aging and some chronic diseases.

Weight Management and Appetite Control

The fiber found in figs provides a feeling of fullness, which can help control appetite and maintain a healthy weight.

Digestive Support

Figs are known for their mild laxative properties, helping to prevent constipation while soothing the digestive system.

Blood Sugar Regulation

Figs have a relatively low glycemic index, which means they release sugar more slowly into the bloodstream, helping to regulate blood sugar levels.

Cardiovascular Health

Figs contain heart-beneficial compounds, such as potassium, which may help regulate blood pressure, and fiber, which may lower cholesterol levels.

Bone Strengthening

Calcium and potassium found in figs support bone health, preventing osteoporosis and fractures.

Use in Medical Cooking

The fig can be integrated into modern medical cuisine in a multitude of ways. It can be eaten fresh as a snack, added to cereal, smoothies or salads, or even cooked into healthy main dishes or desserts.

Medical Cuisine and Taste Pleasure

One of the most appealing features of modern medical cuisine is its balance between health and delight. The fig, with its natural sweetness, richness and variety of flavors, adds an exquisite taste dimension to medical cuisine, making the quest for health more rewarding.

The fig in modern medical cuisine illustrates nature's astonishing potential to guide our quest for healthy, fulfilling lives. By integrating the nutritional and therapeutic benefits of fig into our daily diet, we enhance our ability to promote health and well-being with every bite. It is an invitation to discover the marriage between succulent flavor and health benefits, a gourmet alliance for a balanced body and a fulfilling life.

Chapter 108: The Art of Growing the Fig Tree in a Pot: Tips and Advice

Growing the fig tree in a pot is an exciting adventure that allows lovers of this delicious plant to let their passion flourish, even in limited spaces. When grown in a pot, the fig tree transforms into a living work of art, providing not only succulent fruit but also a touch of natural elegance to any environment.

Choosing the Right Pot and Location

The choice of pot is crucial. Opt for a sufficiently large pot, at least 40 cm in diameter and depth, to allow roots to develop. Make sure the pot has drainage holes to prevent excess moisture. Place the pot in a sunny location, preferably near a window.

illuminated.

Variety Selection

Some varieties of fig trees are better suited to growing in pots than others. Opt for dwarf or compact varieties that will thrive in a small space.

Substrate and Drainage

Use a well-draining potting mix. Mix in sand or perlite to improve drainage.
This will prevent excessive moisture buildup, which can be detrimental to the roots.

Watering and Fertilization

Water regularly, allowing the soil to dry slightly between waterings. Avoid excess water which can cause root rot. Fertilize with a balanced fertilizer during the growing season, usually spring and summer.

Size and Formation

Pruning the potted fig tree is essential to maintain a compact and manageable shape. Prune dead, diseased or malformed branches, and remove suckers growing at the base of the tree. You can also trim to maintain the desired shape and promote better air circulation.

Winter Protection

If you live in an area with cold winters, protect the potted fig tree by placing it in a sheltered location or insulating it with insulating material. Pollination

If you are growing a potted fig tree indoors, it may be necessary to hand pollinate the flowers using a soft brush to ensure fruit formation.

Pest and Disease Monitoring

Watch for signs of pests and diseases, such as aphids, mealybugs or root rot. Act quickly to prevent their spread.

Harvest and Preservation

Pot-grown figs can be harvested once they are ripe. Pick them gently to avoid damaging the delicate skin. Figs can be eaten fresh, dried or used in various recipes.

Growing a fig tree in a pot is a journey that combines gardening and aesthetics. This is an opportunity to enjoy the delights of the fig, even in limited spaces. With the right care and knowledge, you can create a corner of lush nature where the beauty of the potted fig tree brings a touch of elegance and flavor to your daily life.

Chapter 109: Fig Trees in Community Gardens: Cultivating Conviviality and Sustainability

Community gardens are oases of sharing, connection and sustainability in the heart of cities. Among the jewels of these green living spaces, fig trees stand as symbols of connection with nature and shared generosity. Fig trees in community gardens transcend simple fruit trees to become essential parts of a thriving and fulfilling community.

Culture and Education

Fig trees in community gardens provide a unique opportunity to educate community members about plant cultivation, biodiversity and growth cycles. These living trees become natural classrooms where people of all ages can learn together.

Nourish the Body and Mind

Fig trees offer an abundance of sweet and nutritious fruits. They are a source of healthy and delicious food for community members, strengthening local food security.

Strengthening Social Ties

The cultivation and harvest of figs become moments of meeting and exchange within the community. Community gardens, enriched by fig trees, create a space where residents bond with each other around nature and the generosity of the earth.

Ecological Sustainability

Fig trees, with their ability to grow in diverse environments, can play a crucial role in the ecological regeneration of urban spaces. Their leaves, branches and fruits contribute to the nutrient cycle and local biodiversity.

Creation of Meditation Spaces

Fig trees, with their elegant branches and lush leaves, provide ideal shaded spaces for meditation, relaxation and contemplation amidst the hustle and bustle of the city.

Promote Community Engagement

The presence of fig trees in community gardens can encourage more community members to get involved and participate in the management of these spaces. This reinforces the sense of belonging and local pride.

Connecting Generations

Fig trees have the ability to bring together different generations around a common activity. Elders share their knowledge of fig cultivation with young people, creating a living cultural heritage.

Promote Health and Well-Being

The presence of fig trees in community gardens encourages healthy eating and active living.

Fig picking and tree tending become practices that promote physical and emotional health.

Fig trees in community gardens are more than just fruit trees. They embody the values of sharing, sustainability, connection and kindness within a community. These trees, silent witnesses of exchanges and collective growth, weave links between people and nature, thus contributing to the creation of a common space where everyone can flourish and prosper.

Chapter 110: The Art of Greenhouse Cultivation for Fig Trees: A Setting for Controlled Growth

Greenhouse growing provides a controlled environment where nature and science combine to promote optimal plant growth. Among the treasures of the greenhouse, the fig tree stands as a magnificent example of this harmonious marriage.

Benefits of Greenhouse Cultivation for Fig Trees
• **Protection against External Conditions:**Greenhouse fig trees are sheltered from bad weather, temperature fluctuations and strong winds, creating a stable environment conducive to growth.

Extension of the Growing Season: Greenhouses allow •
extend the growing season, providing the opportunity to harvest figs for longer.

Environmental Control: Temperature, humidity and•
Light exposure can be carefully adjusted in a greenhouse, providing optimal conditions for fig trees.

• **Protection against Pests:**Greenhouse fig trees are less prone to pests and diseases, reducing the need to use pesticides.

• **Improvement in Fruit Quality:**The controlled environment allows the figs to grow more evenly and enjoy better flavor.

Greenhouse Cultivation Techniques for Fig Trees

Choice of the Greenhouse: Opt for a well-designed greenhouse with systems•
ventilation, shading and heating to regulate the internal environment.

• **Variety Selection:**Choose varieties of fig trees that thrive in greenhouse conditions, usually dwarf or compact varieties.

• **Soil preparation:**Use a well-draining, nutrient-enriched substrate to provide the roots with optimal conditions.

• **Watering and Fertilization:**Be sure to water regularly, avoiding excess water, and fertilize as needed by the plant.

• **Pollination:**If the greenhouse prevents access by natural pollinators, hand pollination may be necessary to ensure fruit formation.

• **Size and Formation:**Prune fig trees to maintain a manageable shape and promote optimal air circulation.

Precautions to take

• **Light Control:**Make sure fig trees get enough natural light, but avoid excess that could burn the leaves.

• **Adequate ventilation:**Good ventilation prevents the accumulation of excessive humidity and reduces the risk of fungal diseases.

• **Rigorous Monitoring:**Monitor fig trees regularly for signs of pests or disease and act quickly if necessary.

Greenhouse cultivation for fig trees is a feat of harmony between science and nature. It is an invitation to create a controlled ecosystem where fig trees can grow and flourish with vigor renewed. Through meticulous attention to detail, in-depth knowledge of the plant's needs and cutting-edge technology, greenhouse fig trees become splendid examples of what the union between man and nature can accomplish for cultivate beauty and flavor.

Chapter 111: The Fig and Sustainable Agricultural Practices: A Fruitful Alliance for the Earth and **Humanity**

Sustainable agricultural practices have become an imperative necessity to preserve our planet and ensure global food security. At the heart of this quest for sustainability, the fig tree presents itself as an eloquent example of harmonious coexistence between agriculture and nature. Fig cultivation and sustainable agricultural practices converge to form a successful alliance for the health of the earth and humanity.

Conservation of Biodiversity

Fig trees, with their many varieties, play a crucial role in preserving agricultural biodiversity. By growing different varieties of fig trees, farmers help maintain a range of unique plants and preserve local ecosystems.

Rational Use of Resources

Sustainable agricultural practices emphasize the rational use of natural resources, including water. Fig trees, with their ability to tolerate drought conditions, can be grown in areas where water is scarce, contributing to more efficient use of water resources.

Reduction of Carbon Emissions

Fig cultivation generally requires less intensive mechanization, thereby reducing carbon emissions from the use of agricultural machinery. Fig trees encourage simpler, more environmentally friendly agricultural practices.

Natural Fertilization

Fig trees, thanks to their nutrient-rich leaves, can be used for natural soil fertilization. By using fallen leaves as mulch or incorporating them into the soil, farmers improve soil fertility in an environmentally friendly way.

Biological control

Fig trees are home to a variety of beneficial insects and organisms that can help control agricultural pests. By encouraging species diversity in and around fig trees, farmers are adopting biological control methods to keep pest populations under control.

Soil Conservation Practices

Growing fig trees often promotes agricultural practices that preserve soil quality.
Deep rooting of fig trees can prevent soil erosion, protecting long-term fertility.

Local Economy and Rural Communities

Fig cultivation can play a vital role in strengthening local economies, providing employment and encouraging local produce. Fig trees on sustainable farms help create prosperous and resilient rural communities.

The fig, a symbol of sustainability and generosity, fits harmoniously into sustainable agricultural practices. Its ability to withstand environmental challenges and deliver nutritious fruits makes it a valuable partner for tomorrow's agriculture. By merging traditional knowledge with modern innovations, fig cultivation and sustainable agricultural practices join forces to nourish the land, the soul and future generations.

Chapter 112: Propagation of Fig Trees by Cuttings: The Power of Regeneration

Vegetative

Cuttings, this ancient method of vegetative propagation, is an art that allows gardeners and farmers to create new plants using parts of a mother plant. Among the trees that lend themselves wonderfully to this technique, the fig tree emerges like a sparkling star, offering a royal road to propagation.

The Science of Cutting Fig Trees

Fig tree cuttings are a relatively simple technique, but they require a careful understanding of the fundamentals. In general, cuttings involve taking a section of a growing branch, cultivating it under optimal conditions and encouraging it to root to give rise to a new plant.

The choice of cuttings

Fig tree cuttings can be taken from young shoots in spring or summer. It is best to choose healthy, non-diseased and well-developed cuttings to ensure optimal success.

Preparing Cuttings

Cuttings should be cut with clean, sharp tools to minimize injury. They should be between 15 and 30 centimeters long and cut at an angle just below a knot.

Rooting Stimulation

Before planting the cuttings, it is recommended to dip them in a rooting hormone to encourage root development. Then, they can be planted in a well-draining substrate.

Optimal Growth Conditions

The cuttings should be placed in a bright location, but not in direct sunlight, to avoid dehydration. A high humidity level around the cuttings also promotes their rooting.

Encourage Root Growth

Generally, roots from fig tree cuttings may appear after a few weeks to a few months. During this period, it is crucial to maintain regular watering, without drowning the young plants.

Transplantation and Continuing Care

Once the cuttings have developed a sufficient root system, they can be transplanted into larger pots or directly into the ground, depending on where they will be grown long term.

Fig tree cuttings are an exciting way to create new life from old ones. This technique, which is based on the power of vegetative regeneration of fig trees, offers gardeners and nature lovers an opportunity to actively participate in the multiplication of these magnificent trees. By mastering the cutting stages, we continue to perpetuate the beauty and richness of these emblematic trees, thus allowing fig trees to prosper and shine in new horizons.

Chapter 113: Fig Trees in Indigenous Cultures of Oceania: The Deep Roots of the

Natural Connection

The islands dotting the vast blue Pacific Ocean are home to indigenous cultures rich in tradition, spirituality and deep connections to nature. Among the elements that have intertwined harmoniously with these cultures, fig trees emerge as guardians of the earth and symbols of the connection between man and the island ecosystem.

Spiritual Meaning

Fig trees often occupy a central place in the myths and beliefs of Oceanic cultures. They are revered as sacred trees, considered the guardians of life and fertility. The fig tree is often associated with gods, spirits or ancestors, representing a divine presence that watches over communities.

Essential Foods

Figs provide a valuable food source in Oceanic regions, where resource availability may be limited. The juicy, sweet fruits are often eaten fresh or dried, providing a nutrient- and energy-rich diet.

The Moreton Fig Tree: An Ecosystem in Itself

The Moreton Fig Tree (Ficus macrophylla), iconic in Australia, illustrates how fig trees can create unique ecosystems. The aerial roots of this giant fig tree form an intricate network that is home to a variety of creatures, epiphytic plants and insects. This remarkable tree embodies the symbiosis between fig trees and their environment, a harmonious balance that characterizes the indigenous cultures of Oceania.

Conviviality and Gathering

Fig trees, often with their large, shady branches, naturally become gathering places for communities. Under their benevolent shadow, people come together to share stories, to celebrate, to meditate and to build social connections.

Crafts and Materials

Fig trees provide useful materials for traditional crafts. Fibers from the roots can be woven to create baskets and ropes, while the wood can be carved to make various utilitarian and decorative items.

The Sustainability of Fig Tree Cultivation

Although modern influences can sometimes transform traditions, the cultivation of fig trees remains ingrained in the hearts and minds of the indigenous people of Oceania. Respect for these iconic trees and the teachings passed down from generation to generation ensure that fig trees will continue to play a significant role in the cultures, spirituality and way of life of Pacific communities.

Fig trees, with their abundant leaves, delicious fruit and deep connection to nature, embody the spirit and soul of Oceania's indigenous cultures. These majestic trees transcend time, symbolizing the continuity of traditions and the harmonious relationship between people and their environment.

In island Oceania, fig trees are more than just trees – they are guardians of the past, allies of the present and promises for the future.

Chapter 114: The Fig and the Harvest Celebration Rituals: A Feast for the Senses and the Spirit

Harvests, symbols of fertility and the abundance of the earth, have been celebrated throughout human history. Among the jewels that the earth generously offers, the fig emerges as a star during harvest ceremonies. Its sweet fruits, rich in flavor and symbolism, have long been essential elements of harvest celebration rituals. The fig transforms into an icon of festivities, awakening the senses and binding communities together through rituals that honor the land and the bounty it offers.

Seasonal Feasts

Harvest celebration rituals mark the different seasons of the year and are intrinsically linked to agricultural cycles. The fig, with its bountiful and seasonal harvest, is often associated with summer and fall feasts, providing a delectable feast for hungry palates.

The Symbolism of Abundance

Figs, with their fleshy and deliciously sweet interiors, symbolize abundance and fertility. Their shape evokes roundness and fullness, echoing the blessings of the generous earth. When figs are presented at the heart of harvest rituals, they embody gratitude towards nourishing earth.

Exchange and Sharing

Harvest celebration rituals are not only gastronomic events, but also moments of community sharing. Figs, often picked in abundance, are distributed among community members, strengthening social bonds and symbolizing solidarity between individuals.

Cultural Rites and Celebrations

Figs, often associated with specific customs and beliefs, can vary in their role within harvest celebration rituals from culture to culture. Some cultures use figs as offerings to the gods as a sign of gratitude, while others incorporate them into ritual dances or traditional games.

Preparation and Ritual Cooking

Figs, fresh or dried, can be prepared in a variety of ways during harvest ceremonies. Fig dishes are often made with care, incorporating symbolic and traditional ingredients. These dishes, prepared with love and dedication, become symbols of cultural attachment and collective celebration.

Connections with the Earth and Nature

Harvest celebration rituals with figs at the center strengthen the connections between communities and the land that nourishes them. They remind us of the importance of sustainable agriculture and the preservation of nature to guarantee future harvests.

Figs, true jewels of nature, become ambassadors of harvest celebration rituals. With their exquisite flavor and deep symbolism, figs unite people around a table full of meaning, traditions and festivities. They remind us that harvests go well beyond simply gathering food; they embody gratitude, sharing and the vital connection between man and the earth.

Chapter 115: Fig Tree Cultivation in a Tropical Climate: Navigating the Warm Breezes of the Prosperity

Tropical climates, with their heat and humidity, create environments conducive to lush biodiversity. At the heart of these dynamic ecosystems is the fig tree, an iconic tree that finds fertile ground in these conditions.

Adaptation to Tropical Climates

Fig trees, native to subtropical and tropical regions, are at home in warm, humid climates. Their lush leaves and ability to tolerate high temperatures make them natural residents of these regions.

The Humidity Challenge

Humidity, characteristic of tropical climates, can be a double-edged sword for fig trees. On the one hand, it promotes rapid and luxuriant growth, but on the other, it can also create a favorable environment for fungal diseases. Good air circulation and proper spacing between trees can help alleviate these problems.

Watering Management

Although fig trees appreciate moisture, it is important not to overwater them to avoid root rot. Moderate and regular watering is generally recommended.

The Choice of Adapted Varieties

In tropical climates, some varieties of fig trees are better suited than others. Varieties that have natural resistance to fungal diseases and have an ability to produce fruit in humid conditions will be more likely to thrive.

Protection Against Diseases

Tropical climates can encourage the development of fungal diseases, such as rust and mold. Preventative treatments, such as using natural fungicides, can help maintain the health of fig trees.

Regular Size

Regular pruning is important to control excessive growth of fig trees in tropical climates. This not only helps maintain their shape, but also promotes better air circulation, reducing the risk of disease.

Generous Harvests

Fig trees grown in tropical climates are often generous in harvests. Their rapid growth and rate of fruiting allow gardeners to harvest abundant fruit for themselves and to share with the community.

Fig cultivation in tropical climates is an exciting endeavor that requires a keen understanding of the interaction between the tree and its environment. By navigating the challenges of humidity, heat and disease, gardeners can create oases of lush greenery and harvest succulent fruit. Fig trees, with their thick leaves and sun-drenched fruits, become symbols of the abundance and vitality that characterize tropical climates, while providing a deep connection between man and nature in these lands blessed by the sun.

Chapter 116: The Propagation of Fig Trees by Layering: An Ancient Method for Cultivating the Natural Connection

Plant propagation has been a central concern of agriculture and gardening for millennia. Among the techniques that have stood the test of time, layering emerges as a reliable and ingenious method for propagating fig trees. This technique, which involves creating new plants from branches of the mother tree, has the power to create genetic continuity while celebrating relationship intimate between man and nature.

An Ancient and Proven Technique

Layering is a venerable propagation technique used since time immemorial. It involves encouraging a branch from a parent tree to develop roots while remaining attached to the original plant. Once the roots are sufficiently developed, the branch can be separated and planted as a new independent plant.

The Steps of Layering Fig Trees

Layering fig trees follows several steps. A chosen branch is lightly incised or barked, stimulating the formation of roots. This incised area is then wrapped in a damp substrate and held in place with a material like plastic or wire. Once the roots are well developed, the new plant is carefully detached and transplanted.

Aerial Layering

Air layering is a commonly used method for fig trees because it allows you to create a new plant without moving the branch from where it is growing. This method is particularly useful for fig trees that are already well established and difficult to move.

The Deep Connection with Nature

Layering fig trees embodies a deep connection with nature and an understanding of the natural processes of growth and reproduction. It reflects how humans can work in harmony with plants, encouraging their intrinsic capacity to regenerate and multiply.

The Preservation of Ancient Varieties

Layering is also a valuable method for preserving old and rare varieties of fig trees. By multiplying these trees by layering, gardeners contribute to maintaining genetic diversity and save valuable species that might otherwise disappear.

A Lesson in Patience and Connection

The process of layering fig trees takes time and patience. It is a reminder that nature follows its own rhythm and that the connections we form with it require constant attention and deep respect.

Layering fig trees is much more than just a propagation technique. It is a celebration of the relationship between man and nature, a method of preserving genetic wealth and a way of honoring the cycles of growth. Through layering, we honor the wisdom of ancient gardeners and their intimate understanding of the magic of nature.

Chapter 117: Fig Trees in Historic Gardens: Witnesses to History Cultivated with Care

Historic gardens are timeless jewels that carry within them the imprints of the past, the stories of previous generations and the eternal beauty of domesticated nature. Among the plant elements that have adorned these enchanting spaces, the fig trees stand like silent guardians of time.

Witnesses to History

The fig trees planted in the historic gardens have witnessed various eras, from the excitement of antiquity to the industrial and cultural revolutions. Their remarkable longevity has allowed them to cross the centuries, carrying within them the memories of bygone eras.

Connections between the Past and the Present

Fig trees planted in historic gardens are much more than just trees. They connect past generations to current generations, weaving a continuous thread of human connection with nature across the centuries. Their presence evokes continuity, a feeling of constancy in a constantly changing world.

Ancient Varieties

Many historic gardens are home to ancient varieties of fig trees, some dating back hundreds of years. These varieties, often heritage, have been cherished and carefully preserved, because they have become tangible links with the past.

Conservation of Rare Species

Fig trees in historic gardens play an important role in the conservation of rare and endangered species. Their seeds and cuttings are sometimes used to preserve unique varieties that might otherwise be lost.

Care and Attention

Gardeners of historic gardens, aware of the historical value of their fig trees, provide meticulous care and attention to these trees. Special pruning techniques, disease treatments, and specific preservation methods are often used to preserve the vitality of these ancient trees.

Artistic Inspiration

Fig trees, with their sculptural forms and majestic branches, have often inspired artists and garden designers throughout the ages. Their charismatic presence adds an artistic dimension to historic gardens, creating striking visual compositions.

Reflection on Time

Fig trees in historic gardens are constant reminders of the passage of time. They evoke a temporal depth and a history that extends far beyond our own experience.

Fig trees in historic gardens are living symbols of history, perseverance and enduring beauty. Their presence testifies to the symbiosis between man and nature, to the capacity of nature à transcending generations and how historic gardens are more than just physical spaces, but living cultural legacies.

Chapter 118: Fig and Asian Culinary Traditions: An Exquisite Fusion of Flavors and Heritage

Asian culinary traditions, rich in diversity and history, are a treasure trove of creativity and taste harmony. Among the many ingredients that have found their way into these cuisines, the fig emerges as a rare nugget, adding a sweet and sumptuous note to the array of Asian flavors.

The Fig in Asian Cuisine: A Gourmet Discovery

The introduction of the fig into Asian culinary traditions is a story of discovery and adaptation. Although the fig is not native to Asia, it has been welcomed with open arms and transformed into a delectable source of inspiration.

The Fusion of Flavors

Asian cuisines are renowned for their skill in blending diverse ingredients to create complex, balanced flavors. The fig adds a sweet and delicate note to these compositions, creating a harmonious fusion with ingredients such as spices, herbs and sauces.

The Role of the Fig in the Kitchen

The fig is used in a variety of ways in Asian cuisine. It can be incorporated into sweet and savory dishes, such as curries, salads, desserts and marinades. Its natural sweetness makes it an excellent addition to sour or spicy dishes.

Gourmet Desserts

In many Asian traditions, the fig is a key element in desserts. It can be made into jams, pastries, ice creams and sweet soups, adding a touch of sophistication for meal.

The Symbolism of the Fig

The fig, with its graceful form and alluring color, is often associated with beauty and abundance in Asian traditions. Its presence in dishes can bring deeper meaning to meals, symbolizing prosperity and bliss.

Modernity and Tradition

The fig has managed to find its way into modern Asian cuisines while respecting ancient culinary traditions. It is loved for its ability to evoke a sense of nostalgia while offering new and innovative flavor combinations.

The fig, with its sumptuous sweetness and versatility, has gracefully integrated itself into Asian culinary traditions. She added a new and exciting dimension to meals, while respecting the depth of Asian history and culture. The fig embodies the spirit of innovation while honoring the foundations of Asian gastronomy, creating a taste experience that subtly marries the past and the present.

Chapter 119: Innovative Grafting Practices for Fig Trees: Cultivating Creativity in
Nature

Grafting, an ancient and essential technique in horticulture, has evolved over the centuries to become a canvas on which gardeners paint their most daring ideas. Among the trees that have benefited from these innovative grafting practices, fig trees stand out for their adaptability and their ability to blend with a multitude of other plant species. In this chapter we explore innovative grafting techniques for fig trees, their role in expanding horticultural possibilities, and their contribution to botanical diversity.

The Renewal of the Registry

Gardeners and researchers have been attracted by the potential of grafting to create new varieties and forms of trees. Fig trees, with their robust character and genetic flexibility, lend themselves perfectly to these experiments.

Grafting Fruit Varieties

One of the most popular grafting practices for fig trees is fruit variety grafting. This involves grafting a variety of fig tree that produces tasty fruit onto rootstock that is disease-resistant or has specific characteristics. This combines the best of both worlds: an exquisite fruit variety with improved growth and resistance qualities.

The Emblem Registry

Bud grafting is a technique which consists of taking a bud from a chosen variety of fig tree and à insert it into an incision made on the rootstock. This method allows specific characteristics of a variety to be quickly propagated while preserving genetic integrity.

The Crown Graft

Crown grafting, also known as T-grafting, is used to fuse two plants together. This technique can be used to combine different species of fig trees, creating unique shapes and unexpected flavor combinations.

The Art of Hybridization

Innovative grafting practices for fig trees have paved the way for experimental hybridization. Gardeners have the opportunity to blend the characteristics of different fig species to create unique, resilient specimens suited to specific environments or particular taste needs.

Botanical Diversity

Innovative grafting practices have helped enrich the botanical diversity of fig trees. By creating new varieties and promoting hybridization, gardeners help preserve the genetic richness of fig trees and prepare these trees to adapt to future challenges.

Innovative grafting practices for fig trees are a testament to man's creativity and collaboration with nature to create horticultural wonders. These techniques allow us to explore new possibilities, merge species to create new ones and preserve botanical diversity in a constantly changing world. Thanks to innovative grafting, fig trees continue to thrive, adapt and inspire a new generation of gardeners and nature lovers.

Chapter 120: Growing Figs in Poor Soil: The Art of Abundant Creation

Growing the fig tree in poor soil is a remarkable demonstration of nature's ability to adapt and thrive in seemingly harsh conditions. Fig trees, renowned for their resilience, have found a way to transform constraints into opportunities, producing sweet and abundant fruits even in less fertile soils. Let's look at tips and strategies that allow gardeners to successfully grow fig trees in poor soils, while celebrating the tenacity and beauty of nature.

The Elegance of Resilience

The fig tree, a symbol of endurance and adaptation, is well adapted to growing in poor soil. Its ability to draw on available resources and adapt to environmental conditions makes it ideal for gardeners who want to make the most of less fertile soils.

Choosing Suitable Varieties

Choosing fig tree varieties suited to poor soils is a crucial step for success. Some varieties are more tolerant of poor soils and can thrive even with limited resources.

Improve Soil Structure

Although soil may be low in nutrients, providing it with adequate structure is essential. Adding organic matter, such as compost or manure, can improve water retention and air circulation, thereby promoting the growth of fig trees.

Water Management

Water management is crucial when growing fig trees in poor soil. Regular and adequate watering is essential to help the roots draw necessary nutrients from the soil. However, it is important not to overwater, as soggy soil can lead to root rot problems.

Nutrient Intake

Although the soil may be low in nutrients, it is possible to provide additional nutrients to fig trees. Using natural, balanced fertilizers can help compensate for the lack of nutrients in the soil.

The Right Size

Judicious pruning of fig trees in poor soil can promote better growth. Regular pruning helps remove dead or diseased branches and focuses resources on healthy parts of the tree.

The Reward of Patience

Growing fig trees in poor soil may require patience, as growth may be slower compared to more fertile conditions. However, gardeners will eventually reap the rewards of their hard work in the form of tasty and healthy figs.

Growing the fig tree in poor soil is a lesson in humility and trust in nature. Fig trees, with their determination to grow and thrive despite challenges, remind us of the beauty and resilience of life. Growing fig trees in less favorable conditions requires a careful and deliberate approach, but it offers exceptional gratification in terms of succulent fruits and a deeper connection with the earth that nourishes them.

Chapter 121: The Multiplication of Fig Trees by Seedling: Sowing the Roots of Abundance

Propagating fig trees by seed is a method that paves the way for the growth of a new generation of trees, while celebrating the cycle of plant life. While other propagation methods, such as cuttings and grafting, are more commonly used for fig trees, sowing offers a unique experience that allows you to closely follow the germination and growth process. Let's see the stages of sowing fig trees, its advantages and the essential considerations for successfully using this propagation method.

The Magic of Sowing

Sowing fig trees is an invitation to dive into the world of germination and growth. This method tracks the seed's journey from its dormant state to its transformation into a flourishing tree.

Harvesting Seeds

The first step of sowing is to collect fig seeds. The seeds can be extracted from ripe figs and cleaned carefully to remove any pulp.

Seed Stratification

Some varieties of fig trees require a stratification period, meaning they must be exposed to cool temperatures for a period of time to break dormancy. This can be achieved by placing the seeds in the refrigerator for a few weeks.

The Sowing

The stratified seeds are then sown in prepared pots or beds with a light substrate and well drained. The seeds are covered with a thin layer of soil and watered gently.

Patience and Observation

Sowing fig trees requires patience and careful observation. Gardeners should monitor seed germination and seedling growth.

Transplantation

Once the seedlings have reached a suitable size, they can be transplanted to permanent locations. When transplanting, it is essential to handle the roots carefully to avoid damage.

The Benefits of Sowing

Propagating fig trees by seed allows the unique genetic characteristics of the parent trees to be preserved. It also provides an opportunity to experiment and explore different varieties of fig trees.

Climate Considerations

It is important to take into account the climatic conditions of your region when sowing fig trees. Some varieties may adapt better to specific climates, which can influence the choice of seeds to sow.

Sowing fig trees is an adventure that offers a fascinating perspective on the plant growth process. This method allows gardeners to connect in a deeper way with the cycle of plant life and appreciate the stages of germination, growth and transformation. By propagating fig trees by seed, we celebrate the diversity of nature while helping to preserve these precious and delicious trees for generations to come.

Chapter 122: Fig Trees in Modern Zen Gardens: A Harmony Between Nature and

Spirituality

Modern Zen gardens, heirs to the centuries-old tradition of Japanese gardens, embody a refined aesthetic and a deep spiritual connection with nature. Incorporating fig trees into these spaces is a reflection of the close relationship between man and nature, as well as a way to create a calming atmosphere conducive to contemplation and meditation.

Modern Zen Gardens: A Contemporary Balance

Modern Zen gardens thrive in urban environments where the hectic pace of contemporary life can exhaust our soul. Inspired by Zen philosophy, they invite tranquility, contemplation and immersion in the present moment.

Fig Trees: A Bridge To Nature

Introducing fig trees into modern Zen gardens creates a tangible connection with nature. Fig trees, with their spreading branches and distinctive leaves, bring a calming, organic touch to these thoughtfully designed spaces.

The Symbolism of Fig Trees

In spiritual and cultural traditions, the fig tree is often associated with wisdom, knowledge and stability. The presence of fig trees in modern Zen gardens can embody these qualities, inviting visitors to connect with their own inner wisdom.

The Meditation Tree

Fig trees provide a quiet refuge for meditation. Their generous shade and delicate leaves create a space for deep reflection, allowing visitors to retreat from the hustle and bustle of the outside world.

The Diversity of Varieties

Fig trees come in a variety of shapes and sizes, making it possible to create unique arrangements in modern Zen gardens. From dwarf fig trees in pots to those stretching gracefully along stone paths, each variety makes its own contribution to the overall aesthetic.

The Patina of Time

Fig trees, with their slow growth and evocative shapes, can add a touch of maturity to a modern Zen garden. Their presence reminds us that beauty is often shaped by time and patience.

Mediation between Man and Nature

Fig trees in modern Zen gardens demonstrate the intimate relationship between man and nature. By integrating them into these sacred spaces, designers and visitors recognize that peaceful contemplation and meditation are bridges to a deeper connection with the natural world around us.

The presence of fig trees in modern Zen gardens offers an opportunity to explore the duality between inner tranquility and outer expression. These magnificent trees are living reminders of our search for meaning and our aspiration for harmony. By flourishing in these spaces of tranquility, fig trees transcend the role of simple vegetation to become symbols of the relationship between man and nature, evoking a visual poetry that speaks directly to the soul .

Chapter 123: Fig in Traditional Beauty Recipes: A Symphony of Food and care

For centuries, figs have been revered not only for their sweet and juicy flavor, but also for their benefits for skin and hair. Ancient civilizations harnessed the nourishing and revitalizing properties of this iconic fruit to create traditional beauty recipes. The intimate relationship between the fig and beauty treatments highlights recipes that have been passed down from generation to generation.

A Reserve of Natural Nutrients

Figs are rich in essential vitamins, minerals and antioxidants that nourish and protect the skin. Their vitamin C content stimulates the production of collagen, improving the elasticity and firmness of the skin.

An Elixir for the Skin

Figs can be made into masks, exfoliants and toners to revitalize the skin. A mask made from fig puree combined with honey deeply hydrates, while a scrub made from figs and sugar gently removes dead cells, revealing radiant skin.

Natural Hair Gloss

Figs are not only beneficial for the skin, but also for the hair. Hair masks based on figs and essential oils strengthen hair, promote growth and add shine.

Holistic Balance

Beauty recipes using figs embody the balance between nature and science. The natural properties of figs combine with traditional wisdom to provide a holistic approach to beauty care.

Cultural Heritage

Figs have played a central role in the beauty traditions of many cultures. From bath rituals to face masks, figs have been used as key ingredients in skincare recipes passed down from generation to generation.

The Power of Ancient Wisdom

Traditional beauty recipes using figs demonstrate the power of ancient wisdom and knowledge passed down from generation to generation. Ancient civilizations understood the importance of nature in maintaining beauty and health.

Adaptation to Modernity

Fig-based beauty recipes continue to evolve to meet modern needs. Commercial products incorporate the benefits of figs into sophisticated formulations, providing a convenient alternative to homemade recipes.

The fig, this ancient and revered fruit, invites us to dive into the world of traditional beauty treatments. By exploring recipes that have been cherished for centuries, we embrace the harmony between man and nature, between food and care. Figs, with their nourishing and regenerative properties, embody a timeless link between the beauty rituals of the past and the current needs of skin and hair. Through these recipes, we inherit a legacy of holistic care, remembering that beauty emanates from nature and that traditional remedies remain among the best beauty secrets that time has never been able to erase.

Chapter 124: Cultivation of the Fig Tree in the Mediterranean Region: An Eternal Dance with the Sun and the sea

The Mediterranean region, with its sun-drenched landscapes and soothing sea breezes, is the cradle of an ancient fig tree culture. This iconic fruit, which has woven its history with Mediterranean peoples since time immemorial, has become a living symbol of the close relationship between man and nature in this region. Here let's see the many facets of fig tree cultivation in the Mediterranean region, its cultural significance, its traditional cultivation methods and its legacy that continues through generations.

The Sweet Embrace of the Mediterranean

The Mediterranean region offers an ideal climate for growing fig trees. Hot, dry summers, mild winters and sea breezes create an environment conducive to figs flowering and ripening.

The Mediterranean Fig Tree: A Tree of Life

The fig tree is deeply rooted in Mediterranean cultures. It is celebrated in myths, cuisine, art and tradition. Fig trees stand as silent guardians of Mediterranean history, testifying to the intimate connection between man and the earth.

Traditional Cultivation Methods

Fig tree cultivation in the Mediterranean region is often based on traditional methods passed down from generation to generation. Fig trees are sometimes planted in rocky or sandy soils, and are drought tolerant once established.

The Symbiosis between the Fig Tree and the Earth

Fig tree cultivation in the Mediterranean region goes beyond pure agriculture. It is a symbiosis between man, earth and nature. Fig trees enrich the soil and create a microcosm rich in biodiversity.

The Mediterranean Feast

Figs are an integral part of the Mediterranean diet. They are available in sweet and savory dishes, in preserves and jams, demonstrating their versatility and gastronomic value.

A Heritage That Resists Time

The fig trees that dot Mediterranean landscapes are often century-old trees. They carry within them history and collective memory, and witness the resilience of Mediterranean communities in the face of the challenges of time.

The Breath of Tradition and Innovation

Although fig tree cultivation in the Mediterranean region is based on centuries-old traditions, it adapts also to modern changes. New methods of cultivation, conservation and marketing have emerged while preserving the unique character of fig cultivation.

Fig tree cultivation in the Mediterranean region transcends the simple act of growing a fruit tree. It is a celebration of history, land and life that intertwine to create a rich and vibrant fabric. Mediterranean fig trees are more than fruit trees; they are guardians of cultural heritage, silent witnesses to the cycles of nature and living symbols of the symbiosis between man and earth. By cultivating fig trees, Mediterranean people honor an ancestral relationship with nature and continue to bring the echo of the past to the heart of their future.

Chapter 125: The Multiplication of Fig Trees by Root Division: A Way of Growth

Surprising

Fig tree propagation is an ancient art that has evolved over time to include various methods, including root division. This innovative technique offers an intriguing and effective way to propagate these majestic trees, allowing gardeners to explore new avenues for growing and sharing the beauty and delicacy of fig trees. In this chapter, we will delve into the intricacies of propagating fig trees by root division, exploring its process, its benefits and its role in preserving the wealth of fig trees around the world.

A Fertile Approach

Root splitting is a propagation method that involves separating part of the root system of a mature fig tree to create a new plant. This method exploits the natural ability of fig trees to develop adventitious roots from their underground stems.

The Amazing Process

To propagate a fig tree by root division, you must carefully dig up a mature fig tree and separate part of its roots by carefully cutting them. The section of roots thus obtained is then replanted in a new location, where it will develop into a new plant.

Advantages and Benefits

Propagating fig trees by root division has several advantages. It allows the genetic characteristics of the mother plant to be preserved, while ensuring rapid and vigorous growth of the new plant. Additionally, this method can be especially useful for varieties of fig trees that are not easy to propagate by other methods.

Preserving Diversity

The propagation of fig trees by root division plays a crucial role in preserving the diversity of fig varieties. By allowing gardeners to create new plants from mature specimens, this method helps preserve and spread the unique characteristics of each variety, preventing the loss of some rarer and more valuable varieties.

The Marriage of the Ancient and the Modern

Root division unites the ancient and the modern in the art of fig tree propagation. Traditional fig growing techniques are married with contemporary knowledge to create a method that combines ancient wisdom and current innovations.

Propagating fig trees by root division is a captivating method that opens up new perspectives for fig lovers and passionate gardeners. It testifies to the infinite capacity of nature to regenerate and renew itself, while preserving the richness and variety of fig trees around the world. This method reminds us that the art of growing fig trees is constantly evolving, adapting ancient methods to the needs and challenges of the present.

Chapter 126: Fig Trees in Contemporary Botanical Gardens: A Journey Through Heritage and Innovation

Contemporary botanical gardens, true sanctuaries of biodiversity and knowledge, are the modern guardians of the world's flora. At the heart of these lush and educational gardens, fig trees, with their ancient history and their many varieties, find a special place. Fig trees enrich contemporary botanical gardens, preserving their heritage while celebrating innovation in the art of plant conservation.

The Educational Role of Botanical Gardens

Contemporary botanical gardens are centers of environmental awareness and research. They provide visitors with essential information on biodiversity, ecology and conservation. Fig trees, as key elements of Mediterranean biodiversity, offer an exceptional opportunity to learn about the cultural history and biological characteristics of these iconic trees.

Fig Trees as Witnesses to History and Culture

Fig trees are often ambassadors of ancient cultures and stories. In contemporary botanical gardens, they tell captivating stories about migration, human interaction with nature, and the importance of fruit trees in sustenance. Visitors can learn how fig trees have shaped the culinary, medical and spiritual traditions of various regions around the world.

Conservation of Rare Varieties

Contemporary botanical gardens play a vital role in preserving rare and endangered varieties of fig trees. By growing and displaying these varieties under controlled conditions, gardens help prevent their disappearance and maintain the genetic diversity of fig trees for future generations.

The Art of Reproduction and Multiplication

Contemporary botanical gardens are living laboratories where the art of plant reproduction is explored and perfected. Fig trees, with their ability to be propagated by different methods, offer a opportunity to develop advanced propagation techniques that could be applied to other plant species.

Innovation and the Creation of Sanctuaries

Contemporary botanical gardens not only preserve the past, they also anticipate the future. Some gardens integrate agroforestry, permaculture and sustainable management techniques to create balanced ecosystems where fig trees coexist with other plants and organisms. This holistic approach promotes the creation of dynamic and resilient biodiversity sanctuaries.

Fig trees, with their rich cultural heritage and biological diversity, have found a valuable place in contemporary botanical gardens. They transcend geographic and cultural boundaries to blend into the living fabric of these spaces of knowledge and preservation. In botanical gardens, fig trees are not just fruit trees, but keepers of history, catalysts of curiosity, and symbols of commitment to conservation and environmental education.

Chapter 127: The Fig Tree and Shamanic Medicinal Uses: A Journey Between Nature and Spirit

Shamanic practices, anchored in the ancestral wisdom of indigenous cultures, weave an intimate link between man and nature, between the material and the spiritual. At the heart of these rituals and ceremonies is the fig, a sacred tree whose medicinal and symbolic properties are integrated into shamanic traditions around the world. The fig is used in shamanic medicinal contexts, embracing both physical benefits and deep spiritual connections.

The Fig Tree: A Bridge Between Worlds

The fig has been revered in many cultures as a sacred tree, symbolizing fertility, life and knowledge. In shamanic traditions, the fig often acts as a bridge between the material and spiritual worlds. Its delicious fruits and medicinal leaves embody the duality of nature, serving as both food and medicine.

The Medicinal Properties of the Fig Tree

Figs are rich in essential nutrients such as fiber, vitamins and minerals. In shamanic practices, they are used to strengthen the body and support digestive health. Fig leaves also have medicinal properties, often used to treat conditions such as diabetes, hypertension and inflammation.

Spiritual and Emotional Healing

In shamanic ceremonies, the fig is often associated with spiritual and emotional healing. It is considered a remedy to balance energies, heal emotional wounds and facilitate the release of past traumas. Some shamans use figs as meditation and focus tools, creating space for introspection and inner healing.

The Ritual and the Connection with the Divine

The fig is often used in shamanic rituals to establish a connection with the divine and the spirit world. Figs are offered as acts of devotion and gratitude to nature. In some cultures, the fig is considered a symbol of spiritual opening, helping shamanic practitioners transcend the limitations of the material world to access deeper levels of consciousness.

The Importance of Respect and Responsibility

Shamanic medicinal uses of the fig are imbued with respect for nature and responsibility to ancestral teachings. Shamans and traditional healers honor the fig by following specific ritual protocols and recognizing the sacred role of the plant in their practice.

The fig, with its combination of medicinal properties and spiritual symbolism, fits seamlessly into shamanic practices around the world. It embodies the deep connection between man and nature, between the earthly and the transcendent. In the shamanic medicinal uses of the fig, we find a powerful reminder that healing and spirituality are closely linked, and that nature is an invaluable source of wisdom and support for those who are open to its teaching.

Chapter 128: Fig Tree Cultivation in Temperate Climates: The Art of Adaptation and Fruitful Harvest

Fig trees, emblematic of the Mediterranean regions, have managed to conquer temperate climates thanks to à their remarkable adaptability. Growing fig trees in environments where winters can be harsh requires an accurate understanding of the tree's needs and appropriate protection techniques. The challenges and strategies of growing fig trees in temperate climates, and horticulture enthusiasts have turned these challenges into a rewarding experience of harvesting sweet delights.

Adaptation of Varieties to Temperate Climate

The first step in successfully growing fig trees in a temperate climate is the selection of suitable varieties. Some varieties, called "hardy fig trees," are specially developed to tolerate colder temperatures. These varieties are chosen for their ability to withstand winter frosts and produce satisfactory fruit despite the shorter seasons.

The Role of Winter Protection

Fig trees in temperate climates often require protection from freezing temperatures. Techniques include wrapping branches in insulating materials, mulching the soil to conserve heat, and even growing in pots to make moving indoors easier during the colder winter months. cold. These strategies allow fig trees to survive the harsh winter and come back strong in spring.

Pruning Pruning for Abundant Harvest

Pruning plays a crucial role in growing fig trees in temperate climates. Cautious pruning, generally carried out at the end of winter, promotes better air circulation, reduces the risk of diseases and facilitates fruit growth. Specific pruning techniques, such as removing damaged or misdirected branches, help create a strong structure for the fig tree.

The Use of Microclimates

Microclimates, which result from garden layout, surrounding structures and plant arrangement, can play a crucial role in successful fig tree cultivation in temperate climates. Planting fig trees near walls or buildings that store and release heat can increase the chances of winter survival and fruit production.

The Reward of Patience and Perseverance

Growing fig trees in temperate climates requires patience and perseverance. Fig trees often take longer to establish roots and begin producing fruit in these environments. However, when horticultural enthusiasts succeed in overcoming climatic challenges and creating the optimal conditions, they are rewarded with delicious, sweet figs that bear the signature of the temperate climate.

Fig growing in temperate climates is a blend of art and science, understanding the needs of the plant and creative adaptation to local conditions. Gardeners who embark on this adventure discover that fig trees, although not natural inhabitants of these climates, can thrive with proper care. Growing fig trees in temperate climates becomes a lesson in patience, careful observation and respect for the whims of nature, while celebrating the delicacy of the succulent fruits that reward this effort.

Chapter 129: The Multiplication of Fig Trees by Shield Grafting: Artistic Fusion and
Generation of Life

Shield grafting is an ancestral technique which creates a harmonious union between different varieties of fig trees. This method of propagation provides an opportunity to preserve specific characteristics while promoting rapid growth and vigor. The art of escutcheon grafting applied to fig trees, the steps, benefits and wonders of this process merge genetic heritage and human expertise.

The Badge: A Form of Plant Art

Bud grafting is often compared to a form of plant art. In this technique, a small piece of a desired variety, called a 'bud', is inserted into a notch made on the rootstock. This artistic union allows the plant to inherit the desired qualities of the crest, creating a kind of plant homage to the beauty and diversity of nature.

The Escutcheon Grafting Process

Bud grafting follows a careful process. A bud is taken from a fig tree with desired characteristics, such as flavor, size or disease resistance. The bud is then inserted under the bark of the rootstock, usually during the period of active growth. Once in place, the patch is attached and sealed with a sealant to promote fusion and prevent infection.

The Advantages of Shield Grafting

Bud grafting has many advantages. It allows rapid multiplication of the chosen varieties, thus preserving the desired characteristics. Additionally, it offers an effective solution for propagating varieties of fig trees that may be difficult to reproduce from seedlings or cuttings. By combining rootstocks adapted to local conditions with selected buds, horticulturists can create robust and productive fig trees.

The Fusion of Two Plant Identities

Bud grafting is much more than just a propagation technique. It's a silent ceremony where two plant identities merge to create a new expression of life. The rootstock provides the necessary structure and vigor, while the escutcheon provides its unique signature and beauty. Together they work in harmony to produce a plant that reflects both tradition and innovation.

The Transcendence of Time and Space

Escutcheon grafting transcends time and space by connecting generations and places. This technique has been practiced for centuries, crossing borders and cultures to ensure the survival and prosperity of the most popular fig varieties. Each grafted bud is a direct link to gardeners of the past, a continuation of a tradition as old as horticulture itself.

Bud grafting is a process that celebrates human creativity and the diversity of nature. Through this technique, gardeners honor the beauty of fig trees and preserve the traits that make them special. Bud grafting is an act of love towards nature, a dialogue between artist and plant, and an opportunity to weave a continuous thread between generations, connecting the past to the present and to the present. ;future.

Chapter 130: Fig Trees in Ecological Gardens: Natural Symbiosis and Sustainability
Verdant

Ecological gardens embody the harmony between man and nature, highlighting environmentally friendly practices and the preservation of biodiversity.

The Symbiosis of Fig Trees in Ecological Gardens

Fig trees, with their lush growth and succulent fruits, make a significant contribution à the ecosystem of an ecological garden. Their leaves provide a source of shade and shelter for various creatures, while their figs attract a variety of animals, including birds, insects and small mammals. This symbiosis helps to strengthen the food chain and promote biodiversity.

Enrichment of Plant Diversity

Integrating fig trees into an ecological garden also strengthens plant diversity. The fig tree, with its different varieties, adds a new and interesting dimension to the palette of plants already present. Its dense foliage and unique growing characteristics provide habitat for a variety of beneficial insects, creating an ecological balance conducive to garden health.

Natural Fertilization and the Life Cycle

Fig trees contribute to soil fertility through the decomposition of their leaves and fallen fruit. This nourishes the soil, releasing essential nutrients for other plants to grow. By providing a source of organic matter, fig trees participate in the natural cycle of life in the ecological garden.

The Water Economy and Resilience

Some fig trees, like the prickly pear (Opuntia), are adapted to arid environments and can survive with little water. By integrating them into an eco-friendly garden, homeowners can save water while adding aesthetic appeal and ecological value. Their ability to withstand harsh conditions also enhances the sustainability of the garden.

Education and Awareness

Fig trees, with their rich cultural history and ecological contribution, can also be used as educational tools. They provide an opportunity for garden visitors to experience botanical diversity and learn about local ecosystems. Fig trees thus become ambassadors of sustainability and environmental preservation.

Fig trees, with their vital role in creating sustainable ecosystems, are becoming key players in ecological gardens. Their presence promotes biodiversity, supports soil fertility and strengthens resilience to environmental changes. By integrating these iconic trees into ecological gardens, ecology enthusiasts create spaces where natural beauty, biological diversity and harmonious coexistence come together, celebrating a vision of sustainability and respect for the Earth.

Chapter 131: The Fig and Native American Healing Practices: Ancient Link between Nature and Health

Native American people have long enjoyed a deep relationship with nature, using plants and natural resources to maintain their health and well-being. Among these resources, the fig has held a special place as a source of food, traditional medicine and spiritual symbol. The fig has been integrated into Native American healing practices, illustrating the deep connection between indigenous traditions and nature.

Fig as Food and Medicine

For Native American people, the fig was not only a delicious fruit, but also a source of essential nutrients. Figs provided vitamins, minerals and fiber needed for a balanced diet. Additionally, they were used for medicinal purposes. Figs have been recognized for their digestive properties, anti-inflammatory effects and ability to support the immune system.

The Fig Tree in Native American Spirituality

The fig tree also held an important place in Native American spirituality. Some tribes considered the tree sacred and revered it for its strength and vitality. Fig trees were sometimes used as landmarks in the spiritual landscape and were associated with stories and ritual ceremonies. The fig tree embodied the deep connection between human beings, the earth and the cosmos.

Healing Practices

The leaves, roots and fruits of the fig tree were used in various healing practices. Native Americans used the leaves to prepare infusions, believing in their healing properties to relieve digestive ailments, inflammations and even respiratory problems. The roots were sometimes transformed into poultices to soothe muscle pain. Fresh figs were also considered a beneficial food for the body.

The Transmission of Knowledge

Knowledge of the fig's medicinal properties and healing methods was passed down from generation to generation within Native American communities. The elders shared their knowledge with the younger ones, thus guaranteeing the preservation of these traditional practices. This oral and practical transmission helped to keep indigenous customs and beliefs alive.

Contemporary Resonance

Today, while Native American traditions are maintained and respected, the fig continues to play a role in certain healing practices among indigenous communities. Although access to traditional resources may vary, the fig continues to be a powerful reminder of ancient wisdom and nature's ability to support human health and well-being.

The fig, with its dual nature as food and medicine, played an important role in the healing practices of Native American peoples. She embodies the deep relationship between indigenous traditions and nature, illustrating how plants can be essential allies in the pursuit of health and well-being. By exploring the role of the fig in Native American healing practices, we celebrate the richness of indigenous wisdom and the power of nature as a healer.

Chapter 132: Fig Tree Cultivation in Restricted Urban Areas: The Art of Getting the Best from One **Limited Space**

In urban areas where space is often a luxury, growing fig trees can seem like a challenge. However, with the right knowledge and creative techniques, it is possible to grow these iconic fruit trees even in tight spaces. Let's look at strategies and tips for successfully growing fig trees in limited urban environments, and how this practice contributes to the enrichment of urban life.

Choosing Suitable Varieties

When growing fig trees in restricted urban areas, the choice of varieties is crucial. Opt for dwarf or compact varieties that are suitable for smaller spaces. Fig trees in pots or grafted onto dwarf rootstocks can be ideal options for growing in containers on balconies, terraces or even windowsills.

Containers and Espaliers

Using containers suitable for growing fig trees provides valuable flexibility in urban spaces. Potted fig trees can be moved depending on light and seasonal conditions. Additionally, fig trees can be trained into an espalier shape against walls or fences, optimizing the use of vertical space.

Optimal Growth Conditions

Be sure to provide optimal growing conditions for your fig trees in urban areas. Choose sunny locations where trees can receive at least 6 hours of direct sunlight per day. Fig trees need well-drained soil and a balanced diet to thrive. Containers require special attention to watering and fertilizing.

Size and Formation

Pruning is an essential element of fig tree cultivation in restricted urban areas. Fig trees can be pruned to maintain their compact size and encourage a specific shape, such as the espalier. Regular pruning also stimulates fruit production and prevents problems related to crowding.

Winter Protection

In areas with harsh winters, it is important to protect fig trees from cold temperatures.

Potted fig trees can be moved indoors during the winter season. For fig trees in the ground, use thick mulches around the base of the tree to protect the roots from frost.

Benefits for Urban Living

Fig tree cultivation in restricted urban areas brings a series of benefits for city life. Besides producing delicious fruits, fig trees add a touch of greenery and beauty to the urban environment. They provide spaces for relaxation and meditation, and encourage connection with nature even in the most densely populated places.

Fig tree cultivation in a restricted urban area may seem demanding, but with the right approach, it is completely achievable. By choosing suitable varieties, using containers and espalier techniques, and providing the necessary care, urban gardening enthusiasts can enjoy the delights of figs even in the smallest spaces. This practice not only brings tasty harvests, but it also adds a natural touch to urban life, strengthening the connection between man and nature in the heart of cities.

Chapter 133: The Propagation of Fig Trees by Split Grafting: The Art of Perpetuating Tradition

Split grafting is a revered propagation technique that allows gardeners to perpetuate their favorite varieties of fig trees while preserving their unique characteristics. This age-old method provides a reliable way to propagate fig trees, while allowing gardeners to create new combinations of roots and grafts.

An Ancient Historical Link

Split grafting has been used for centuries to propagate different species of plants, including fig trees. This technique has been passed down from generation to generation, becoming an integral part of horticultural tradition. It allowed gardeners to preserve and share their favorite varieties, thus guaranteeing the diversity and sustainability of fig trees in gardens.

The Advantages of Split Grafting

Split grafting offers several advantages. It allows gardeners to retain the desirable characteristics of a specific variety, such as size, flavor and disease resistance. Additionally, this technique helps shorten the time needed for a young fig tree to reach maturity and start

à produce fruit. Split grafting also provides precise control over the propagation process, allowing gardeners to choose rootstocks suited to local conditions.

Steps of Split Grafting

1. **Selection of Scions and Rootstock:** Choose a healthy, vigorous rootstock as well as scions from a mature, productive fig tree.

2. **Preparation of the Graft:** Cut a scion from the branch of the mother fig tree, making sure it has several buds.

3. **Preparation of the rootstock:** Make a slit-shaped incision on the rootstock, close to the ground. There slot must be clean and precise. 4. **Insert the Plugin:** Carefully insert the scion into the rootstock slot, making sure the inner layers match.

5. **Ligation and Protection:** Use a ligature to hold the graft in place. Apply putty or grafting tape to protect the grafting area.

6. **Interview :** Place the young scion in optimal growing conditions. Keep the soil moist and avoid excessive stress.

7. **Elimination of Unwanted Shoots:** As the scion grows, remove unwanted shoots that emerge from the rootstock.

The Transmission of Knowledge

Cleft grafting transcends the simple act of multiplication. It represents the transmission of knowledge between generations, with each gardener learning from elders and adding their own experience to the tradition. This technique honors the history of fig trees and contributes to their future, preserving the genetic richness and diversity of varieties.

Split grafting is much more than a technique for propagating fig trees. It is an act of preservation, of passing on tradition and creating the future. Gardeners who master this method honor past generations while leaving their own mark on the fig trees to come. It is a tribute to the symbiosis between nature and the hand of man, which has allowed fig trees to thrive and enrich our gardens and our lives for centuries.

Chapter 134: Fig Trees in Green Roof Gardens: An Elegant Fusion of Nature and
Urbanity

Green roof gardens have gained popularity in densely populated urban areas, providing a green oasis in the heart of urbanity. Incorporating fig trees into these elevated spaces adds a new dimension to this trend, offering not only the alluring aesthetic of fig trees, but also the ecological benefits and natural connection they bring. Fig trees fit harmoniously into green roof gardens, enriching the urban experience while promoting sustainability.

The Rise of Green Roof Gardens

Green roof gardens are much more than aesthetic features. They play a vital role in regulating urban temperature, improving air quality and managing stormwater. They also provide spaces for relaxation, recreation and even gardening, which is essential in urban environments where floor space is limited. The integration of nature at height offers a renewed experience of city life.

The Elegance of Fig Trees

Fig trees bring a timeless Mediterranean touch to green roof gardens. Their leaves

Lush greens create a refreshing and relaxing atmosphere, while juicy fruits add a palette of colors and flavors. Fig trees are also perfect for espalier pruning, making them an ideal choice for maximizing the use of vertical space in elevated gardens.

Environmental Benefits

Incorporating fig trees into green roof gardens brings a range of environmental benefits. Their leaves help with thermal regulation by providing shade and reducing radiant heat. Additionally, fig trees absorb CO2 and emit oxygen, helping to improve air quality.

By adding layers of vegetation, fig trees help filter air pollutants and mitigate the effects of urban heat islands.

Connection with Nature at Height

Green roof gardens offer residents precious moments of connection with nature right in the heart of the city. The presence of fig trees adds an organic and lively dimension to this space, encouraging city dwellers to connect with the natural cycles of growth and harvest. The ability to grow fig trees on rooftops brings food production closer to the consumer, promoting an increased appreciation of where food comes from.

Conservation of Urban Biodiversity

The integration of fig trees into green roof gardens contributes to the conservation of biodiversity in urban areas. Fig trees provide habitat for various species of insects, birds and other small creatures, thereby strengthening the ecological balance of the urban environment. Green roof gardens with fig trees thus become havens of biodiversity in concrete deserts.

Fig trees, with their timeless elegance and environmental benefits, find their natural place in green roof gardens. By integrating these iconic fruit trees into elevated urban spaces, we strengthen the connection between nature and city life. Fig trees bring

Mediterranean touch and a vital connection with the earth, transforming rooftops into green havens and reservoirs of sustainability.

Chapter 135: Fig in Traditional Asian Remedies: A Treasure of Benefits for the
Health

For millennia, traditional Asian remedies have drawn on the riches of nature to promote health and well-being. The fig, with its various medicinal properties, occupied a special place in these ancestral practices.

A Source of Essential Nutrients

Figs have long been recognized in traditional Asian remedies for their nutritional value. Rich in fiber, minerals such as potassium, calcium and magnesium, as well as vitamins, figs are a valuable source of essential nutrients for the body. These nutrients support heart health, digestion and bone health, making them a key component of many natural remedies.

Yin and Yang balance

In traditional Chinese medicine, the balance between Yin and Yang is fundamental to maintaining health. Figs, with their sweet, fresh flavor, are often considered Yin, meaning they have a cooling effect on the body. They are used to counter excess internal heat, calm agitation and soothe skin irritations. Figs are also associated with the kidney and spleen in Chinese medicine, supporting digestive and kidney health.

Fig in Ayurveda

In Ayurveda, India's traditional system of medicine, figs are recognized for their cooling and calming qualities. They are used to reduce heat and inflammation in the body, especially during hot times of the year. Figs are also considered a tonic for the digestive system and are often recommended to treat constipation problems.

A Remedy for Respiratory Ailments

In many Asian cultures, figs have been used to treat respiratory ailments such as coughs and respiratory tract infections. Fig syrup is made by simmering figs in water with honey or sugar, then straining the liquid. This syrup is often used as a natural remedy to soothe a sore throat and relieve coughs.

Antioxidants and Immune Health

Figs are rich in antioxidants such as polyphenols, which help neutralize free radicals and protect cells from oxidative damage. This antioxidant capacity strengthens the immune system and can help prevent chronic diseases. Figs have also been shown to have anti-inflammatory properties, making them beneficial for various inflammatory conditions.

The fig, with its richness in nutrients, its Yin and Yang balance, and its various medicinal benefits, plays a significant role in traditional Asian remedies. Various cultures have incorporated the fig into their healing practices, recognizing its calming, antioxidant and nutritional properties. As a natural health treasure, the fig continues to shine in traditional Asian remedies, offering its valuable benefits to current and future generations.

Chapter 136: Fig Cultivation in Drylands: An Odyssey of Resilience and Survival

Fig cultivation in drylands is a captivating example of how nature and man can work together to create a lasting symbiosis. Fig trees, with their ability to thrive in harsh conditions, embody the resilience of plant life in the face of climatic adversity. This chapter explores the art and science of fig cultivation in drylands, highlighting the unique challenges, innovative strategies, and rewards that come with it.

Adaptation to Arid Constraints

Fig trees have evolved to adapt to arid environments, developing physiological and morphological characteristics that allow them to survive in drought conditions. Their dense, glossy foliage reduces water loss through evaporation, while their deep roots explore water tables for moisture. These adaptations allow them to withstand the rigors of the arid climate, making them symbols of resistance and tenacity.

Water Management

In arid areas, water management is essential for growing successful fig trees. Drip irrigation systems, use of recycled water and rainwater harvesting are vital techniques for maintaining soil moisture. Traditional methods, such as building retention ponds to collect rainwater, are often used to maximize irrigation efficiency and reduce waste.

Fertilization and Soil Amendments

Soil fertility in drylands can be a major challenge. Fig trees benefit from well-drained soils, rich in organic matter and nutrients. Adding compost, manure and organic materials helps
à improve soil structure and retain moisture. Balanced fertilization practices, tailored to the specific needs of the fig tree, are essential to support the healthy growth of the tree.

Protection Against Climate Extremes

Dryland fig trees face climatic extremes, such as high temperatures during the day and significant drops at night. Planting fig trees near buildings or structures can provide some protection from wind and create more favorable microclimates. Using organic mulch around trees helps retain soil moisture and protect roots from extreme temperatures.

Rewards of Fig Cultivation in Arid Zones

Growing fig trees in drylands presents significant rewards. Fig trees provide a valuable food source in areas where food resources may be limited. Their dense shade provides refuge from the scorching sun, creating pleasant spaces for relaxation and socializing. In addition, fig trees contribute to the regeneration of arid ecosystems by improving soil quality and promoting biodiversity.

Fig cultivation in drylands is a living illustration of nature's ability to adapt and human ingenuity to take advantage of local resources. Resilient fig trees embody perseverance in the face of climatic challenges, while offering ecological and nutritional benefits. In a world facing climate change and increasing environmental stresses, the art of growing fig trees in drylands reminds us of the importance of collaboration between man and nature to create a future sustainable.

Chapter 137: The Multiplication of Fig Trees by Branch Grafting: The Art of Perpetuation

Vegetable

Propagating fig trees by branch grafting is a traditional and proven technique that preserves the desirable characteristics of a particular fig tree while accelerating its growth and spread. This expert method, a blend of skill and science, reveals man's ingenuity in imitating the natural processes of plant reproduction to create strong, productive trees.

The Science of Grafting by Rameau

Twig grafting, also known as English-style grafting, involves the union of a twig of the desired fig variety (called the scion) onto a compatible rootstock. The success of this technique depends on the precise alignment of the vascular tissues between the scion and the rootstock, which allows nutrients and water to flow efficiently.

Selection of Scions and Rootstocks

The key to successful branch grafting lies in the judicious choice of scions and rootstocks. THE

Scions are taken from mature, healthy fig trees, ideally during their dormant period. The rootstocks, which can be seedlings or wild fig plants, must be compatible with the graft variety. This compatibility guarantees good contact and harmonious growth.

Grafting Steps

The twig grafting procedure generally follows these steps: First, a precise cut is made on the rootstock, creating a flat surface. Next, a graft is taken, cut at an angle to maximize the contact surface. The two sections are assembled so that the cambiums (vascular layers) coincide, ensuring a continuous flow of sap.

Types of Branch Grafting

There are several methods of branch grafting, including split grafting, bud grafting, and crown grafting. Each of these techniques has its advantages and limitations, but they all aim to achieve a strong union between the scion and the rootstock.

Results and Benefits

Branch grafting allows you to quickly multiply fig trees with specific characteristics, such as fruit flavor, disease resistance or vigorous growth. This method also promotes faster growth and early fruit production compared to growing from seed. By preserving the genetic characteristics of a valuable variety, branch grafting contributes to the diversity of cultivated fig trees.

Plant Heritage

Branch grafting transcends time and connects generations by allowing the transmission of precious trees from one generation to the next. Special varieties and grafting skills are preserved, ensuring that trees with exceptional qualities will continue to thrive. This process of perpetuation plant represents a form of cultural and natural heritage that links gardeners, orchardists and nature lovers to an enriching past.

Branch grafting for fig tree propagation is an expression of human ingenuity and love for nature. By combining botanical knowledge with technical precision, this method allows for the rapid and efficient propagation of desired fig varieties. By celebrating and preserving these exceptional trees, branch grafting ensures the sustainability of unique fruit trees and contributes to the richness of the plant heritage.

Chapter 138: Fig and Natural Animal Care: A Beneficial Alliance

For centuries, the fig has been recognized for its nutritional and medicinal benefits for humans. However, its beneficial influence also extends to the animal kingdom. From food to natural remedies, figs have found their way into care for pets and livestock.

Healthy and Natural Food

Figs make a nutritious and tasty addition to animal diets. Rich in fiber, vitamins and minerals, they offer benefits for digestion, the immune system and general health. Giving figs to pets, such as dogs and horses, can contribute to a varied and balanced diet.

Natural Remedies for Animal Health

Figs are also used in natural veterinary medicine. For example, dried figs can be used to treat constipation in pets. Their fiber content promotes regular intestinal transit. The antioxidant properties of figs can also help boost the immune systems of animals, just as they do for humans.

Prevention and Management of Skin Problems

Fig has soothing and anti-inflammatory properties that can benefit pets' skin. Fig preparations can be used to soothe itching, irritation and rashes in pets with dermatological problems. The natural enzymes found in figs can help in the healing of minor wounds.

Holistic Care for Farm Animals

Farmers and ranchers also recognize the benefits of figs for livestock. Figs can be used as dietary supplements to improve the general health of livestock. The digestive and anti-inflammatory properties of figs may help maintain intestinal health in livestock, reducing the need for chemical medications.

Respect for the Environment

The use of figs in natural care for animals is part of an environmentally friendly approach. Natural and organic methods for animal health minimize the use of harmful chemicals, which benefits both animals and the surrounding ecosystem.

The fig is a prime example of how nature provides an abundance of beneficial resources for animal health and well-being. From nutritional supplements to skin and digestive remedies, the fig has proven its potential in natural care for pets and livestock. By integrating this delicious and nourishing treat into pet care, owners and breeders take a holistic approach that promotes long-term health and vitality.

Chapter 139: Fig Trees in Educational Gardens: Cultivating Knowledge Through the Seasons

Educational gardens play a vital role in learning and raising awareness about the environment, sustainable agriculture and nature. Among the various plants that find their place in these gardens, the fig tree occupies a particularly enriching position. Growing fig trees in educational gardens goes beyond just growing plants – it is an opportunity to teach future generations the value of nature, the beauty of botanical diversity and the deep connections between man and the earth.

Lessons on Biology and Ecology

Growing fig trees in educational gardens provides a unique opportunity to teach students about plant biology, including pollination, growth, propagation and life cycles. The fig trees are particularly suited to illustrating the concepts of ecological interactions, mutualism between plants and pollinators, and the dependence of many animals on the fruits produced by fig trees.

Practical Learning of Sustainable Agriculture

Fig trees, being hardy and undemanding in care, are ideal for introducing students to sustainable agriculture. By involving students in planting, pruning, watering and harvesting fig trees, educators can teach the principles of natural resource management, biodiversity preservation and responsible use of fig trees. water and nutrients.

Discovery of Culture and History

Growing fig trees in educational gardens also provides the opportunity to explore the cultural and historical aspects of this plant. Students can learn how fig trees have been grown and used in different societies throughout time. They can learn about the culinary traditions, customs and myths associated with fig trees in various cultures around the world.

Healthy Eating Awareness

Figs, rich in nutrients, fiber and antioxidants, can be used to educate students about the importance of a healthy and balanced diet. By integrating figs into nutrition education programs, teachers can show students how food choices can influence their health and well-being.

Promote Appreciation of Nature

Growing fig trees in educational gardens allows students to connect more deeply with nature. By observing the growth of fig trees, admiring their leaves and fruit, students develop a deeper understanding and appreciation of the beauty and complexity of the natural world around them.

Fig trees in educational gardens embody the perfect fusion of hands-on learning, environmental awareness and cultural discovery. By providing students with opportunities to grow, observe and interact with these exceptional trees, educators open doors to enriching and engaging learning. The lessons learned from growing fig trees in educational gardens go far beyond horticultural skills – they shape the minds and hearts of students as responsible stewards of the planet, knowledge and the beauty of nature.

Chapter 140: Fig and Biological Pest Control Techniques: An Approach

Natural for Crop Protection

The fight against parasites and pests is one of the major challenges of modern agriculture. While many farmers are turning to chemical methods to protect their crops, an alternative that is more respectful of the environment and human health is emerging: biological control. In this context, the fig tree stands out as a valuable ally thanks to its intrinsic properties which promote the natural regulation of parasite populations. The relationship between the fig and biological control techniques highlight the benefits of this approach for agricultural sustainability.

Promoting Biodiversity and Ecological Balance

Fig trees, as a natural habitat and food resource for a variety of animals, attract a multitude of species. Birds, bats and predatory insects find refuge in fig trees, creating a diverse ecosystem that promotes the natural regulation of pest populations. Fig trees act as "hotels" biological, attracting parasite predators, which helps maintain an ecological balance in the fields.

Attractive to Natural Predators

Figs produce volatile compounds that attract predatory insects such as parasitoid wasps. These wasps are known to parasitize the larvae of many agricultural pests, such as aphids and caterpillars. By growing fig trees near vulnerable crops, farmers can encourage the presence of these wasps, thereby reducing pest pressure.

Create Refuge Areas for Predators

Fig trees can also serve as refuge areas for natural predators. These trees provide shelter and breeding grounds for beneficial insects that feed on pests. Fig trees thus act as biological oases within crops, promoting the reproduction and preservation of natural predators.

Reducing Pesticide Use

Using biological control techniques involving fig trees can reduce reliance on chemical pesticides. By encouraging the presence of natural predators, farmers can maintain pest populations at acceptable levels without resorting to chemicals potentially harmful to the environment and human health.

Education and Awareness

The combination of growing fig trees and using biological control techniques can also be an educational opportunity for farmers. By learning to observe the interactions between fig trees, pests and predators, farmers can gain a deeper understanding of agricultural ecology and ways to promote crop health sustainably.

The fig, with its properties that encourage biodiversity and the presence of natural predators, fits perfectly in biological control approaches against parasites. By using fig trees as a tool to create balanced ecosystems in fields, farmers can reduce their reliance on chemical pesticides while maintaining crop and environmental health. This alliance between the fig and biological control techniques illustrates the importance of working with nature to guarantee sustainable and resilient agriculture.

Chapter 141: Fig Tree Cultivation in Urban Vegetable Gardens: Elevating Nature in the Heart of the City

À As urban spaces develop and intensify, urban vegetable gardens become true oases of greenery within the urban environment. Among the plants that find their place in these urban gardens, the fig tree shines with its ability to bring a Mediterranean touch and a tasty harvest to these restricted spaces. Growing fig trees in an urban vegetable garden is not only limited to the practical aspect of food production, but also strengthens the connection between city dwellers and the

nature, while inviting a piece of the countryside into the heart of the city.

The Urban Adaptability of the Fig Tree

The fig tree is a plant perfectly suited to growing in an urban vegetable garden. Its slow growth and moderate size make it an ideal choice for small spaces. With proper care, a fig tree can thrive in a pot, producing lush leaves and succulent fruit, bringing a touch of greenery to the urban environment.

Promoting Urban Biodiversity

Fig tree cultivation in urban vegetable gardens offers a valuable opportunity to promote biodiversity in urban areas. Fig trees attract a variety of pollinating insects, birds and other small animals, creating a mini-ecosystem in the heart of the city. This biodiversity contributes to the ecological balance and resilience of urban ecosystems.

Environmental Education and Awareness

The presence of fig trees in urban vegetable gardens can also play an essential role in environmental education. City dwellers, by observing the life cycle of fig trees, from flowering to harvest, can gain a deeper understanding of natural processes. This promotes awareness of nature and encourages a more respectful attitude towards the environment.

Community Engagement and Well-Being

Growing fig trees in urban vegetable gardens can strengthen the sense of community among local residents. Shared gardening spaces provide a gathering place where people can interact, exchange knowledge and share harvests. Actively participating in fig cultivation can also improve mental well-being by providing city dwellers with a peaceful escape from the urban frenzy.

Challenges and Rewards

However, it should be noted that growing fig trees in an urban vegetable garden can present challenges. Regular care, adequate watering and protection against pests are aspects to take into account to guarantee the success of the crop. Despite this, the rewards are numerous. The satisfaction of harvesting fresh figs in an urban environment, the ornamental beauty of the trees, and the opportunity to educate and raise awareness in the community are well worth it.

Growing the fig tree in an urban vegetable garden transcends the simple act of growing plants. It is a statement in favor of nature in the midst of urbanization, an opportunity to bring city dwellers closer to the natural environment and a way to beautify urban spaces. By placing fig trees in urban vegetable gardens, we celebrate the intersection between nature and urban life, reminding city dwellers of the richness and beauty of the natural world, even in the heart of the city.

Chapter 142: Fig and Soil Conservation Methods: Cultivate to Protect

Soil conservation is one of the fundamental pillars of sustainable agriculture. Healthy and fertile soils are essential to maintain agricultural productivity, preserve biodiversity and mitigate the effects of climate change. In this quest to protect and restore our precious soil resources, the fig tree emerges as an unexpected but powerful ally.

Strong Roots and Extended Root System

Growing the fig tree has unique soil conservation benefits due to its deep and extensive root system. The roots of the fig tree help stabilize soils and prevent erosion, especially in areas prone to heavy rain or strong winds. These deep roots also play a crucial role in preventing soil depletion, as they draw nutrients from deep down and bring them to the surface.

Protection Against Erosion

Fig trees, planted in rows or hedges, can serve as natural barriers against erosion. Their large, dense leaves create a ground cover that slows water runoff and reduces the risk of soil erosion. By protecting the soil against erosion, fig trees help maintain the structure of the soil and preserve its fertility.

Improvement of Soil Organic Matter

The leaves that fall from fig trees, rich in nutrients, decompose and form a layer of organic matter on the soil. This organic matter helps improve soil structure, increase its water-holding capacity and promote beneficial microbial activity. By naturally nourishing the soil, fig trees support the overall health of the agricultural ecosystem.

Beneficial Association with Other Cultures

The practice of intercropping fig trees with other crops can also contribute to soil conservation. Fig trees, as perennial trees, can create more stable microclimates, reducing water evaporation from the soil. They can also play a role as a windbreak, protecting crops sensitive to strong winds and reducing soil moisture loss.

The fig tree, with its robust nature and powerful root system, proves to be a key player in soil preservation. In an era where fertile soils are threatened by urbanization, intensive agriculture and climate change, integrating fig trees into agricultural systems can offer valuable solutions for soil conservation. Fig trees thus illustrate how the symbiosis between plants and the earth can create a more sustainable and resilient future for our agriculture and our environment.

Chapter 143: Seasonal Care for a Healthy Fig Tree: Nourish, Protect and Cultivate

Growing and maintaining a healthy fig tree requires constant attention throughout the year, adapted to seasonal changes. From spring pruning to winter protection, each season brings its own demands for maintaining the health and vitality of this exceptional fruit tree.

Spring: The Time for Growth

Spring marks the start of a new growing season for the fig tree. Now is the time to prune dead, diseased or damaged branches to encourage vigorous growth. Also prune crossing branches to allow better circulation of air and light. During this period, the application of a balanced fertilizer promotes the development of new buds and the formation of fruits.

Summer: Blossoming and Harvest

In summer, the fig tree enters its flowering and fruiting phase. During this period, it is important to maintain regular watering to prevent the soil from drying out too much, which can lead to premature fruit drop. Adding a layer of mulch around the base of the fig tree helps conserve moisture and reduce competition from weeds.

Autumn: Preparing for Winter

À As fall approaches, the fig tree begins to slow down its growth. This is the time to stop applying fertilizer to avoid a late growth spurt that could be vulnerable to frost. On the other hand, maintaining adequate watering is important until the tree goes dormant. Ripe fruits are harvested gradually, taking care not to damage the branches.

Winter: Protection Against the Cold

Winter is the time when the fig tree goes dormant. In areas with harsh winters, it may be necessary to protect the tree from the cold by wrapping the branches with insulating material or covering it with a tarp. It is also important to monitor soil moisture and ensure that it does not become too dry.

Seasonal care is essential to maintaining the health and productivity of a fig tree throughout the year. By understanding the tree's specific needs in each season, gardeners can promote optimal growth, bountiful harvests and increased resistance to disease and environmental stresses. Caring for a fig tree thoughtfully and consistently not only rewards gardeners with flavorful fruit, but also creates a deep connection with the rhythm of nature and the cycles of plant life.

Chapter 144: Fig Cultivation in Drought Conditions: A Guide to Farming Resilient

Drought has become a major challenge for agriculture in many parts of the world. Faced with increasingly limited water resources and changing climate conditions, farmers are looking for sustainable solutions to grow drought-resistant crops. In this context, the fig tree presents itself as a promising option thanks to its ability to adapt to conditions of low water availability. This chapter explores strategies and practices for growing the fig tree in drought conditions, highlighting its potential to contribute to food security and the resilience of agricultural systems.

Natural Adaptation to Drought

The fig tree, native to arid Mediterranean regions, is naturally adapted to drought conditions. Its thick, fleshy leaves allow it to store water, giving it the ability to withstand periods of lack of humidity. Additionally, its deep and extensive root system allows it to access moisture deep in the soil, providing a vital water source when the surface layers are dry.

Choice of Resistant Varieties

To successfully grow fig trees in drought conditions, choosing suitable varieties is essential. Some varieties are better suited to lower humidity levels than others. Landraces and indigenous varieties may have developed better drought tolerance over time, making them wise choices for growing in arid regions.

Water Management

Effective water management is crucial in fig cultivation under drought conditions. Drip irrigation is a recommended method, as it delivers water directly to the roots, minimizing waste and preventing excessive evaporation. It is also important to mulch the soil around the fig tree to reduce evaporation and conserve moisture.

Soil Preparation and Fertilization

Preparing the soil ahead of time is essential to help fig trees survive and thrive in drought conditions. Enriching soil with organic matter can improve its ability to retain moisture and provide essential nutrients. Slow-release fertilizers or well-rotted composts can feed fig trees on a regular basis without creating an excessive growth spurt.

Growing the fig tree in drought conditions illustrates the plants' ability to adapt and thrive in harsh environments. Through its natural adaptations and management practices appropriate, the fig tree can play an important role in food security and the sustainability of agricultural systems in regions facing drought-related challenges. By promoting crop resilience in the face of climate change and diminishing water resources, fig tree cultivation shows how collaboration between nature and agriculture can provide solutions for a more sustainable future.

Chapter 145: The Fig and the Use of its Leaves as Natural Fertilizer: An Approach

Ecological Fertilization

In the world of sustainable and environmentally friendly agriculture, the use of natural resources to fertilize soil is increasingly encouraged. Fig leaves, often overlooked, hide amazing potential as a source of nutrient-rich natural fertilizer. Fig leaves can be collected, prepared and used as an organic fertilizer to promote soil health and increase crop fertility.

The Nutritional Value of Fig Leaves

Fig leaves, rich in nutrients such as nitrogen, phosphorous and potassium, are a great way to nourish the soil naturally. Nitrogen, in particular, is essential for plant growth and protein formation. The leaves also contain minerals like calcium, magnesium and iron, which are essential for healthy crop growth.

Collection and Preparation of Leaves

To use fig leaves as a natural fertilizer, it is important to collect them correctly. Choose healthy leaves, not damaged by diseases or pests. After the leaves fall in autumn, gather them together and let them dry in the shade. Once dry, the leaves can be shredded to facilitate their incorporation into the soil.

Use as Fertilizer

Fig leaves slowly decompose in the soil, gradually releasing their nutrients over time. They can be used in different ways:

1. **Composting**: Fig leaves can be added to a compost pile to increase its nutrient content. When compost decomposes, it can be added to soils to improve their fertility.

2. **Mulch**: Laying a layer of shredded leaves around plants acts as a natural mulch, helping to retain moisture, prevent weed growth and nourish the soil as the leaves decompose.

3. **Infusion**: Soaking fig leaves in water for several days creates a nutrient-rich infusion. This solution can be used to water plants, providing immediate nourishment.

Ecological and Economic Benefits

Using fig leaves as a natural fertilizer offers several benefits. It reduces dependence on chemical fertilizers, thereby contributing to soil health and environmental preservation. Additionally, it makes use of an often overlooked resource, reducing costs and minimizing agricultural waste.

Harnessing fig leaves as a natural fertilizer is an environmentally friendly approach to enriching soils and promoting crop growth. By adopting this practice, farmers can cultivate their land sustainably, creating a balanced ecosystem where natural waste nourishes the soil, and the soil nourishes the plants. The fig, often celebrated for its delicious fruits, can thus offer a valuable contribution to soil fertility and the prosperity of crops.

Chapter 146: Summer Pruning to Promote Fruiting: Key Element in Fig Tree Management

Pruning is an essential practice in growing fig trees, playing a crucial role in their health, shape and ability to produce abundant fruit. While winter pruning is well known, summer pruning is just as important, particularly when it comes to promoting fruiting. This chapter explores the principles and benefits of summer pruning to encourage fig production, while maintaining the vitality and vigor of the trees.

Summer Pruning: A Specific Role

Summer pruning differs from winter pruning because it primarily aims to manage excess growth and direct the tree's energy toward fruit production. In summer, fig trees tend to develop long, vigorous shoots. By pruning judiciously during this period, one can control the size of the tree and encourage the development of flower buds which will turn into delicious figs.

Summer Pruning Basics

1. **Shoot Thinning**: Remove weak, damaged or misplaced shoots. Focus on removing branches that cross or rub together, which can create disease-prone areas.

2. **Encouragement of Flower Buds**: Identify shoots that are bearing flower buds for the next fruiting season. Avoid excessively pruning these areas, as this may eliminate fruit-producing sites.

3. **Growth Control**: Reduce the length of excessively long shoots to encourage more compact, concentrated growth. This will allow the tree to devote more energy to producing fruit rather than excessive growth.

Benefits of Summer Pruning for Fruiting

1. **Increase in Fruit Production**: By removing unnecessary branches and promoting flower buds, summer pruning creates an environment conducive to abundant fruit production.

2. **Improvement of Fruit Quality**: By focusing the tree's energy on fewer fruits, summer pruning can result in better size and eating quality of figs.

3. **Managing Tree Size**: Summer pruning keeps the tree at a manageable size, preventing overgrowth

uncontrolled which could harm the health of the tree and the ease of harvesting.

Specific Summer Pruning Techniques

1. **Pruning Shoot Tips**: Cut the ends of growing shoots to encourage denser branching and flower bud formation.

2. **Reduction of Shoot Length** : Prune shoots that are too long to encourage growth compact.

3. **Elimination of Secondary Shoots**: Remove small side shoots that form near developing figs, as they can divert energy from fruit growth.

Summer pruning to promote fruiting is an essential practice to maximize fig production and quality. By understanding the basic principles of summer pruning and applying specific techniques, gardeners can not only enjoy more abundant and flavorful harvests, but also maintain the health and shape of their fig trees. Summer pruning is a proactive method for growing balanced and productive fig trees, creating an ideal environment for lovers of these delicious fruits.

Chapter 147: Cultivation of the Fig Tree in Permaculture: A Harmonious Synergy with Nature

Permaculture, an agricultural design approach based on the principles of natural ecosystems, offers an innovative and sustainable vision for growing food.

Promote Diversity

A fundamental pillar of permaculture is diversity. Fig trees add a unique dimension to this diversity thanks to their adaptability to different climates and soils. By placing fig trees wisely within a permacultural ecosystem, we can create microclimates favorable to the growth of other plants and encourage biodiversity.

Optimal Use of Resources

Permaculture values the optimal use of available resources. Fig trees, with their ability to grow in varied soils and their resistance to drought, align perfectly with this philosophy. Their deep roots can also help prevent soil erosion.

Cycle of Resources and Energy

In permaculture, the emphasis is on creating closed cycles of resources and energy. Fig trees, by producing an abundance of leaves, fruit and branches, provide a valuable source of organic matter for composting and fertilizing. Fig trees can also be a source of shade for other plants, helping to regulate soil temperature and preserve humidity.

Promoting Resilience

Permaculture aims to create resilient systems capable of coping with climate change and disruption. Fig trees, thanks to their ability to adapt to varied conditions, contribute to the resilience of the ecosystem. By promoting crop diversity and including fig trees, we create a more robust and resilient ecosystem.

Cooperation with Wildlife

Fig trees are often pollinated by insects and birds, making them valuable for permaculture systems that seek to encourage beneficial wildlife. Fig trees can also provide food for wildlife, contributing to the local food chain.

Permaculture fig cultivation embodies the spirit of cooperation with nature, while creating sustainable and resilient systems. By integrating fig trees into a permacultural ecosystem, we create a symbiosis between man and nature, where diversity, sustainability and regeneration are the key words. Fig trees are not just fruit trees, but valuable contributors to creating an agricultural world in harmony with natural cycles.

Chapter 148: The Fig and the Benefits of Cultivation in Containers: Savoring Generosity in a Space Restricted

Container growing offers a fascinating opportunity for fig lovers to enjoy the delights of these sweet fruits even in limited spaces.

Optimization of Limited Space

In urban environments or small gardens, space is at a premium. Growing in containers maximizes the use of limited space while providing a touch of lush greenery. Container-grown fig trees can be placed on balconies, decks, or even patios, allowing everyone to participate in the joy of growing and harvesting their own figs.

Mobility and Flexibility

The containers offer the ability to easily move fig trees depending on seasonal needs, optimal sunlight and climatic conditions. This mobility ensures that trees receive the exposure needed for healthy growth, while allowing gardeners to create aesthetically changing configurations in their outdoor space.

Soil Quality Control

Container cultivation allows precise control of soil quality. This is particularly useful in nutrient-poor soils or areas where the soil is not suitable for growing fig trees. By using a high-quality potting soil and compost mix, gardeners can provide fig trees with all the nutrients they need to flower and fruit.

Ease of Disease and Pest Management

Growing in containers allows for increased control of diseases and pests. Fig trees in containers are more insulated from soil pests and pathogens that could otherwise harm their health. This can reduce the need to use pesticides and promote a more environmentally friendly approach.

Aesthetics and Versatility

Container fig trees bring natural beauty and elegance to any space. Their lush leaves and graceful habit add a decorative touch while providing delicious fruit. Additionally, bins can be chosen based on the desired aesthetic, allowing harmonious compositions to be created.

Growing fig trees in containers is a clever and rewarding way to grow these magnificent fruit trees in even the smallest of spaces. It offers the opportunity to enjoy the benefits of figs, their sweet sweetness and their enchanting beauty, while being creative in the arrangement of the outdoor space. The practical, aesthetic and functional benefits of growing in containers make fig trees an attractive choice for anyone who wants to savor nature's bounty, no matter the size of their garden.

Chapter 149: Managing Common Fig Tree Diseases: Preserving a Precious Source of fruits

Growing fig trees can be a rewarding experience, but it can also come with challenges, especially when it comes to diseases that can affect these fruit trees. This chapter examines common fig tree diseases, explores their causes, and offers management strategies to preserve these valuable fruit sources.

Anthracnose: A Fungal Enemy

Anthracnose is one of the most common fungal diseases affecting fig trees. It appears as brown or black spots on leaves, fruits and branches. Wet conditions favor the spread of anthracnose. To manage it, it is essential to maintain good air circulation around the trees, prune infected branches and avoid excess humidity.

Fig Rust: Signs to Watch For

Fig rust is another common disease. It causes orange-colored pustules to appear on the leaves, leading to their discoloration and premature fall. To prevent the spread of rust, it is important to remove and destroy infected leaves as soon as possible. Applying copper-based treatments can also help contain this disease.

Gray Rot: A Wet Threat

Gray rot, caused by the fungus Botrytis cinerea, thrives in humid conditions and can affect ripe figs and vegetative parts of the tree. To prevent it, it is recommended to maintain good ventilation around the trees and avoid excess humidity. Removing infected parts and regularly harvesting figs also helps in the management of this disease.

Bacterial Canker: A Serious Challenge

Bacterial canker is a bacterial disease that causes lesions on the branches and stems of fig trees. Lesions can cause branch death and reduce tree vigor. Prevention involves proper pruning to prevent spread and application of copper-based treatments.

Managing common fig tree diseases is essential to preserving these valuable fruit trees and continuing to enjoy their delicious fruits. By adopting appropriate cultural practices, closely monitoring signs of disease and intervening quickly when necessary, it is possible to minimize the effects of disease on fig trees. Prevention and management strategies, combined with a holistic approach to tree health, will help maintain the beauty and productivity of fig trees in our gardens.

Chapter 150: Cultivation of Fig Trees in Restricted Spaces: A Delight Within Reach

In the modern world where outdoor space is often limited, growing fig trees can seem like a challenge. However, with appropriate methods and careful planning, it is possible to make these majestic fruit trees even in confined spaces. This chapter explores strategies and tips for successfully growing fig trees in compact environments.

Choice of Adapted Varieties

The first step to growing fig trees in tight spaces is to choose varieties suited to that situation. Opt for dwarf or compact varieties that will thrive best in containers or small gardens. Dwarf fig trees generally produce fewer long branches, making them easier to adapt to limited space.

Well-Choose Containers and Bins

Containers and bins are valuable allies when space is limited. Opt for containers suited to the tree's expected adult size. Make sure they have drainage holes to prevent excess moisture. Containers also provide the flexibility to move fig trees depending on light and climate conditions.

Controlled Size and Shape

Judicious pruning of fig trees is essential in tight spaces. By pruning long branches and promoting a compact shape, you encourage vertical growth and minimize horizontal clutter. Regular pruning also helps prevent branches from becoming invasive.

Use of Espaliers and Trellises

Fig trees can be trained into espaliers along walls or trellises, maximizing the use of vertical space. Espaliers not only save space, but they also add a decorative touch to the environment. By shaping them to your needs, you can create artistic shapes while optimizing space.

Optimal Location Selection

Choose the location carefully for your fig trees in tight spaces. Look for an area that receives enough direct sunlight, as fig trees need at least six hours of sunlight per day to produce quality fruit. If possible, keep them away from heavily shaded areas and make sure they are not exposed to strong winds.

Watering and Fertilization Management

Fig trees in tight spaces may be more sensitive to water and nutrient quality. Monitor soil moisture carefully and water appropriately. Use a balanced fertilizer to support growth and fruiting. Container sizes may require more frequent fertilization, so adjust your practices accordingly.

Growing fig trees in tight spaces requires patience, planning and attention to detail. However, the rewards are many: fresh, delicious and fragrant figs at your fingertips, a natural touch in urban environments and the satisfaction of growing an exceptional fruit tree despite space constraints. With the right practices, you can enjoy the splendor of the fig tree even in the limited setting of a yard, patio or small garden, creating a corner of lush nature where joy to cultivate and harvest fruit is fully realized.

chapter 151: The Fig and Beneficial Associations with Other Plants: A Natural Symbiosis

When it comes to gardening, the practice of combining plants is a strategic approach that can improve crop health, stimulate growth and maximize plant utilization. space. In this context, the fig proves to be a valuable ally, offering mutual benefits when combined with certain plants. Beneficial associations with other plants enrich the cultivation of the fig tree.

Companion Flowers: Attracting Pollinators

Integrating companion flowering plants near fig trees can encourage cross-pollination, thereby increasing fruit yield. The flowers attract pollinators such as bees, which play a role essential in fig production. Plants like lavender, rosemary and sages attract bees while providing aesthetic and aromatic benefits.

Ground Cover Plants: Preserving Humidity

Ground cover plants have the ability to maintain soil moisture and reduce competition from weeds. When combined with fig trees, they help keep the soil cool and prevent excessive weed growth which could harm tree growth. Choices such as mint, lemon balm and clovers can be wise.

Companion Vegetables: Maximizing Space

Pairing vegetables with fig trees can be beneficial in maximizing the use of space. Fast-growing vegetables, such as radishes and spinach, can be planted between rows of fig trees to take advantage of available space before the fig trees fully blossom. This allows a double harvest on the same plot.

Repellent Herbs: Keeping Pests Away

Some herbs have natural repellent properties that can help keep potential pests away from fig trees. For example, sage can repel harmful insects while adding an aromatic touch to the garden. This combination can reduce the need to use pesticides while maintaining the health of fig trees.

Companion Trees: Creating a Microclimate

Combining fig trees with other trees can help create a favorable microclimate. Taller trees provide partial shade, which can be beneficial for fig trees as they can avoid stress caused by excessive sun exposure. This approach is especially useful in areas where intense sunlight can affect the growth of fig trees.

Plant association is an ingenious way to take advantage of the natural interactions between plant species. By combining the characteristics and benefits of different plants, a balanced and productive ecosystem can be created. In the case of the fig, beneficial associations with other plants can improve pollination, soil health, pest protection and space utilization. By applying plant association principles, gardeners can not only grow thriving fig trees, but also create diverse and resilient gardens that benefit the entire environment.

Chapter 152: Crop Rotation for Healthy Fig Trees: A Smart Strategy

Crop rotation is an age-old, time-tested practice of planning and rotating crops on a plot of land to optimize plant health, prevent disease, and maintain soil fertility. Although this practice is often associated with annual crops, it can also be adapted for fruit trees such as fig trees.

Diversification and Disease Prevention

Crop rotation for fig trees involves varying the types of plants grown on the same plot. This helps prevent the accumulation of fig-specific pathogens in the soil. Certain diseases, such as root rot or downy mildew problems, can develop if fig trees are grown in the same location for many years. Rotating crops reduces the risk of recurring infections and maintains the overall health of the fig trees.

Optimization of Soil Fertility

Crop rotation also promotes soil fertility. Each type of plant has specific nutritional needs. Alternating crops prevents depletion of the same nutrients in the soil, which can occur if fig trees are continually grown in the same location. Some crops, such as legumes, can even help enrich the soil by fixing atmospheric nitrogen and supplying it to the soil, which indirectly benefits fig trees.

Pest and Disease Reduction

A well-planned crop rotation can also help reduce pressure from pests and diseases specific to fig trees. Insect pests and pathogens that thrive on one type of plant may struggle to survive in the absence of their preferred host. By introducing other crops between the fig trees, we disrupt the life cycle of these harmful organisms and reduce their presence.

Improvement of Soil Structure

Crop rotation can also improve soil structure. Deep-rooted plants can penetrate the soil deep, improving drainage and overall soil structure. This can be particularly beneficial for fig trees, as well-drained soil encourages their growth and reduces the risk of root rot.

Crop Rotation Planning for Fig Trees

Crop rotation for fig trees can be planned over a period of several years. It is important to select crops that are not susceptible to the same diseases as fig trees. Legumes, herbs and fast-growing crops can be great options. By alternating these crops with fig trees, we encourage the diversity and health of the ecosystem.

Crop rotation is a smart strategy for maintaining the health of fig trees and preventing disease and pest problems. By varying the crops on the same plot, we promote soil fertility, reduce the risk of diseases specific to fig trees and improve the structure of the soil. This centuries-old practice is a powerful tool for gardeners concerned about the health of their fig trees, and it can contribute to a bountiful harvest and vigorous growth of fruit trees.

Chapter 153: Fig Cultivation in a Humid Climate: Challenges and Solutions

Growing fig trees in a humid climate presents both unique benefits and challenges. While

Fig trees generally thrive in Mediterranean regions with dry climates, it is entirely possible to grow them successfully in more humid climates by taking specific steps to prevent problems related to excessive humidity.

Humid Climate Challenges

Excessive humidity can cause several problems for fig trees. Wet conditions encourage the growth of fungi, mold and fungal diseases, such as root rot and mildew. Additionally, fig trees are more susceptible to disease in humid climates because humidity encourages the spread of pathogens. The roots of fig trees can also rot due to soil saturation, which can result in limited vegetative development and poor fruiting.

Solutions for Growing Fig Trees in a Humid Climate

1. **Selection of Adapted Varieties:**Opt for varieties of fig trees adapted to humid climates. Some varieties are more resistant to humidity than others. Look for varieties that are less susceptible to fungal diseases and have better moisture tolerance.

2. **Improved Drainage:**Improve soil drainage using methods such as raising the growing bed, adding gravel or sand to the soil, and creating raised mounds. Good drainage prevents water accumulation around the roots.

3. **Choice of Location:**Choose a location where fig trees have good air circulation to reduce stagnant humidity. Avoid low areas where water could collect.

4. **Pruning and Aeration:**Prune the branches of the fig tree to promote better air circulation and greater exposure to sunlight. This will help reduce moisture on the leaves and prevent fungal diseases.

5. **Moderate watering:**Although fig trees love water, overwatering can lead to excessive soil moisture. Water moderately and avoid direct watering on the leaves to reduce the risk of

fungal diseases.

6. **Avoid Over-fertilization:**Excess fertilizer can encourage excessive leaf growth and increase susceptibility to disease. Use balanced fertilizers and follow recommendations to avoid over-fertilization.

7. **Disease Prevention:**Apply preventive treatments against fungal diseases, such as copper or sulfur sprays. Perform regular inspections for signs of disease and act quickly if necessary.

Growing the fig tree in a humid climate can be a challenge, but with the right measures in place, it is entirely possible to achieve a successful harvest of flavorful figs. Selecting suitable varieties, improving drainage, watering management and disease prevention are all key to successfully growing fig trees in a humid environment. By combining these strategies, fig lovers can enjoy growing this exceptional fruit tree even in climates where humidity predominates.

Chapter 154: Fig and Protection from Common Pests: Strategies for Cultivation
Flourishing

Growing fig trees can be a rewarding endeavor, but like any fruit crop, it is prone to pest attack. These small, voracious insects can cause considerable damage to the leaves, flowers and fruit of fig trees. However, by understanding the most common pests and implementing appropriate protection strategies, it is possible to maintain the health of your fig trees and ensure a bountiful harvest.

Common Pests Affecting Fig Trees

1. **Aphids:**These sap-sucking insects can infest the leaves and young shoots of fig trees, causing leaves to curl and the tree to lose vigor.

2. **Fruit Flies:** The females of these flies deposit their eggs in developing fruits, causing brown spots and rotting of the fruit.

3. **Mealybugs:** These small scale-like insects attach themselves to branches and leaves, sucking the sap and weakening the tree. 4. **Apple codling moth:** The larvae of this butterfly burrow into figs, making the fruit unfit for consumption. 5. **Spider mites:** Mites can attack the leaves of fig trees, causing the leaves to turn yellow and wilt.

Protection Strategies

1. **Regular Monitoring:** Inspect your fig trees regularly for signs of pests. Early detection allows rapid intervention.

2. **Growing Companion Plants:** Some plants naturally repel pests. Planting aromatic herbs like mint, rosemary or lavender nearby can help deter pests.

3. **Use of Natural Insecticides:** Choose insecticides based on natural products, such as insecticidal soap or neem oil, to treat light infestations.

4. **Encouraging Natural Predators:** Ladybugs, lacewings and parasitoid wasps are natural predators of pests. Creating an environment conducive to their presence can help maintain ecological balance.

5. **Size of Affected Parts:** If you identify parts of the tree that are severely affected by pests, consider pruning and removing them to prevent spread.

6. **Trapping:** Use sticky traps or pheromone traps to catch pests before that they do not damage your fig trees.

7. **Crop rotation:** If you have several fig trees, try planting them in different locations from one year to the next to avoid the concentration of pests.

Protecting fig trees from pests is essential to ensure a healthy and abundant harvest. By combining regular monitoring, natural prevention methods and targeted intervention in the event of an infestation, it is possible to minimize pest damage and maintain the vigor of your fig trees. By following these strategies, you can enjoy the succulent delights of figs while maintaining the health of your fruit trees.

Chapter 155: The Importance of Pruning for Fig Tree Productivity: Cultivating Results
Successful

Pruning, a practice often considered an art, plays a vital role in the health and productivity of fig trees. These elegant and nourishing fruit trees benefit greatly from careful management of their growth. Pruning is not just a cutting technique, but a discipline that requires a thorough understanding of the specific needs of the fig tree and its growth cycles. As we explore the importance of pruning for fig tree productivity, we discover how this practice can lead to successful results.

1. Stimulate Growth and Fruiting

Strategic pruning encourages the development of new shoots and branches, thereby stimulating the growth of the tree. Pruning back dead, diseased or damaged branches encourages the circulation of air and light within the canopy, allowing fig trees to produce more energy through photosynthesis. This energy is then directed towards the growth of new branches and the formation of flowers and fruits, thereby improving overall productivity.

2. Size and Shape Control

An unpruned fig tree can become bulky and disorganized, which can hinder light penetration and aeration of the tree. Regular pruning maintains the desired size and shape, making fruit harvest and general tree management easier. Proper pruning also prevents the tree from becoming too dense, which would reduce the quantity and quality of figs produced.

3. Elimination of Diseases and Pests

Pruning can help remove parts of the tree affected by disease or invaded by pests. Removing these parts prevents the spread of problems and preserves the overall health of the tree. Additionally, opening the canopy through pruning makes it easier to apply natural or organic treatments to combat infections.

4. Encouragement of Branching and Secondary Branching

Proper pruning encourages branching and secondary branching of branches. This means the tree develops more potential fruiting sites, increasing fruit production capacity. Well-distributed branching also ensures that fruits receive adequate exposure to sunlight, which can improve their flavor and ripening.

5. Tree Resource Management

Pruning allows the tree to manage its resources efficiently. Fig trees have a limited capacity to produce nutrients and water. Pruning helps the tree to concentrate its resources on the most vital parts, such as new shoot growth and fruit production, rather than wasting them on old, unproductive parts.

Meticulous pruning is an essential practice for maximizing the productivity and health of fig trees. By adapting pruning techniques to the specificities of each fig tree, we can promote vigorous growth, abundant fruiting and resistance to disease. For any gardener or grower who aspires to harvest tasty and bountiful figs, careful pruning is a key to growing fruitful results.

Chapter 156: Fig Tree Cultivation in Mediterranean Climate: An Elegant Alliance between Aridity and **Abundance**

The Mediterranean climate, characterized by its hot, dry summers and mild, humid winters, offers a ideal environment for growing fig trees. This harmonious marriage between the specific climatic conditions and the needs of this fruit tree results in a thriving and sustainable crop, marked by an abundance of tasty figs. Let's dive into the details of growing the fig tree in a Mediterranean climate and discover how this elegant alliance between aridity and abundance takes shape.

1. Drought Resistance

Fig trees have evolved a natural adaptation to drought, making them ideal for Mediterranean regions where summers are often marked by intense heat and limited rainfall. Their thick, fleshy leaves reduce water loss through evaporation, while their deep roots allow them to reach moisture reserves deep in the soil. This drought resistance is a valuable asset in a Mediterranean climate where irrigation may be limited.

2. Adaptation to Mild Winters

The mild winters of the Mediterranean climate provide a mild environment for fig trees. These trees tolerate moderately cold temperatures, which means they do not need extreme protection measures during the colder months. However, winters that are too wet can be problematic, as they can cause fungal diseases. Good air circulation and well-drained soil are essential to preventing these problems.

3. Heat Requirement for Fruit Ripening

Figs require a period of sufficient heat to ripen properly. In a Mediterranean climate, the hot summer provides constant and prolonged heat, which stimulates the optimal ripening of figs. High temperatures help develop the sugar content of fruits and contribute to their sweet, delicious flavor.

4. Irrigation Management

Although fig trees are drought tolerant, controlled irrigation is essential to ensure proper good fruit production. During the active growing period, it is best to keep the soil moist but not soggy. Regular irrigation helps ensure that figs ripen evenly and avoid problems such as fruit splitting.

5. Choice of Varieties

In a Mediterranean climate, a well-adapted variety is essential. Drought-tolerant fig varieties with moderate heat requirements and appropriate ripening characteristics are the best candidates. Some popular varieties for Mediterranean regions include 'Noire de Caromb', 'Violette de Sollies', and 'Blanche du Languedoc'.

Fig tree cultivation in a Mediterranean climate is a story of collaboration between the tree and the environment. Fig trees blend harmoniously into the arid landscape, taking advantage of summer heat and drought resistance. The abundance of ripe figs during Mediterranean summers is a reminder of nature's ability to produce delicious harvests even in the hottest conditions. The cultivation of the fig tree in this climate is an ode to the subtle alliance between the plant and its place of growth, creating a nourishing and artistic picture that enriches the Mediterranean region.

Chapter 157: Fig and Methods of Control

Mushrooms: Preserving a Valuable Harvest

The fig, a succulent and nutritious fruit, is a wonder of nature. However, its delicacy makes it prone to attacks by fungi, which can harm its growth and quality. Controlling these fungal invaders is a crucial task to ensure a bountiful and healthy harvest.

1. Cultural Management Practices

The first line of defense against fungus is implementing appropriate cultural management practices. Ensuring adequate air circulation around trees, avoiding planting too densely, helps reduce humidity which encourages fungal growth. Also, the selection of varieties resistant to fungal diseases can reduce the risk of infections.

2. Pruning and Proper Pruning

Regular pruning and trimming dead or diseased branches is essential to prevent fungus from spreading. By removing the affected parts, you limit areas suitable for fungal growth. It is important to disinfect cutting tools between each tree to prevent the spread of spores.

3. Use of Natural Fungicides

Natural fungicides, such as baking soda and neem oil, can be used to prevent and control fungal infections. These substances work by altering the environment favorable to the growth of fungi while minimizing the effects on the surrounding ecosystem.

4. Application of Copper

Copper is an element recognized for its effectiveness in the fight against fungi. Copper treatments, applied judiciously and as recommended, can help control fungal infections. However, excessive use of copper can lead to accumulation in the soil and have negative effects on the environment.

5. Crop Rotation

Crop rotation is an effective strategy for reducing the presence of fungal spores in the soil. Avoiding growing fig trees or other plants susceptible to the same fungi in the same location from year to year can prevent the continued spread of diseases.

6. Regular Monitoring

Vigilance is essential to spot the first signs of fungal infections. Regularly monitoring leaves, fruit and branches for spots, discolorations or other abnormalities can allow for rapid intervention if a problem arises.

Fig cultivation requires special attention to prevent and control fungal infections that can jeopardize the harvest. By combining cultural management practices, natural fungus control methods, and regular monitoring, it is possible to preserve this valuable crop. The fight against fungi in fig cultivation is an expression of respect for this exceptional plant and a guarantee for future generations to enjoy these sweet and nutritious delights.

Chapter 158: Winter Care to Protect the Fig Tree from the Cold: Nourishing Life in the Period of Rest

The fig tree, symbol of generosity and vitality, also goes through periods of winter rest where it needs special care to cope with the cold. Protecting this delicate fruit tree from the harsh winter weather is crucial to ensuring robust health and abundant fruiting in spring.

1. Root Protection

The roots of the fig tree are vulnerable to frost and temperature fluctuations. Applying thick mulch around the base of the tree helps maintain a more stable temperature and prevents damage from repeated freezing and thawing.

2. Reduced Watering

In winter, the fig tree's water requirements are significantly reduced because growth is slowed. Reduce watering to avoid excess moisture around the roots and minimize the risk of root rot.

3. Protection of Buds

The buds of the fig tree are sensitive to intense cold. Wrapping the buds with insulating materials such as straw or non-woven fabrics can help prevent frost damage.

4. Anti-Dehydration

Winter wind and cold can cause excessive moisture loss from leaves. Spraying a fine mist of water on the leaves in dry weather can help reduce dehydration.

5. Lightweight Size

In winter, when the tree is dormant, light pruning can be done to remove dead, diseased or damaged branches. This promotes air circulation and prevents the build-up of moisture conducive to the growth of fungus.

6. Winter Sail

Using specially designed winter sails helps protect the tree while still allowing air circulation. These sails act as a barrier against cold winds and frosts.

7. Early Preparation

Beginning preparations for winter care in the fall allows the tree to gradually acclimate to cold conditions. Gradually reduce watering and apply mulch before freezing temperatures arrive.

Winter care for the fig tree is a vital part of its cultivation. By taking steps to protect the roots, buds and leaves, you ensure the preservation of this majestic tree and its future harvests. This attentive care demonstrates our connection with nature and our commitment to watching over life even in times of apparent rest. By combining traditional knowledge and modern practices, we continue to celebrate the beauty and resilience of the fig tree through the seasons.

Chapter 159: Fig Tree Cultivation in Clay Soils: Transforming Obstacle into Opportunity

Growing fig trees in clay soils may seem like a daunting challenge, but with the right approach and techniques, it is entirely possible to thrive in these conditions.

1. Understanding Clay Soils

Clay soils are characterized by their fine, compact texture, which can result in slow drainage and high compaction potential. However, these soils are rich in nutrients and have the ability to retain moisture.

2. Improvement of Soil Structure

One of the first steps to growing fig trees in clay soils is to improve their structure. This can be achieved by adding compost, decomposed manure and organic materials to increase soil porosity and improve drainage.

3. Creation of Plantation Mounds

To improve drainage, it is advisable to create raised planting mounds. This allows excess water to drain more easily and prevents stagnation around the roots.

4. Selection of Suitable Varieties

Some varieties of fig trees are better suited to clay soils because of their tolerance to moisture and à compaction. It is important to choose varieties that can thrive in these specific conditions.

5. Watering Management

Although clay soils retain moisture, it is essential to manage watering to avoid excess moisture around the roots. Regular, moderate watering is best to avoid root rot.

6. Addition of Organic Materials

By continuing to add organic matter to the soil each year, you help improve its structure over time. This also promotes soil microbial life, which is essential for the overall health of the tree.

7. Raise the Grow Beds

If you are planting in rows or growing beds, elevating these beds can help reduce soil compaction and improve drainage.

8. Avoid Compaction

Avoiding walking or working the soil when it is too wet can prevent excessive compaction of clay soil.

Growing fig trees in clay soils can be a rewarding experience when you consider the specific needs of this fruit tree. By adopting soil improvement techniques, watering management and selection of suitable varieties, you can transform soil considered difficult into an environment conducive to the lush growth and abundant fruiting of fig trees . This is a testament to agriculture's ability to adapt to challenges and take advantage of opportunities to grow rich, nutritious foods, even in less favorable conditions.

Chapter 160:

The Fig and Natural Fertilization Practices: Nourishing the Earth to Nourish the Trees

Fertilization is a crucial component of growing fig trees, as it directly influences the growth, health and fruiting of these delicious fruit trees.

1. Composting: Black Gold for Fig Trees

Composting is a natural fertilization practice that recycles organic waste into a nutrient-rich fertilizer. Kitchen scraps, garden waste and fallen leaves can be turned into high-quality compost, providing fig trees with a continuous source of essential nutrients.

2. Decomposed Manure: A Rich Natural Fertilizer

Decomposed manure from herbivorous animals is a valuable source of organic nutrients for the fig trees. It can be incorporated into the soil to improve its structure and enrich its nutrient content.

3. Organic Mulch: Protect and Nourish

Applying organic mulch around the base of fig trees offers many benefits. In addition to reducing moisture evaporation, mulch gradually breaks down to release nutrients into the soil.

4. Compost Tea: A Nutritious Cocktail

Compost tea is a nutrient-enriched liquid made by steeping compost in water. It can be sprayed on leaves and soil to provide fig trees with additional nutrition.

5. Use of Companion Plants

Certain companion plants can promote the health of fig trees by fixing nitrogen from the air into the soil and repelling pests. Legumes like clovers are good examples.

6. Marine Algae and Rock Flour: Enrich the Soil Naturally

Seaweed and rock flour are rich sources of minerals and trace elements essential for the growth of fig trees. They can be used as soil amendments to enhance its fertility.

7. Crop Rotation: Balancing Nutrients

Rotating crops in your garden can help balance the nutrient needs of fig trees by preventing soil depletion.

8. Avoid Overdose

It is important to note that excessive fertilization can cause nutritional imbalances and damage to fig trees. A measured and careful approach is essential to avoid fertilizer overdoses.

The cultivation of fig trees can be guided by environmentally friendly natural fertilization practices. By adopting methods such as composting, using decomposed manure, organic mulches and natural amendments, you can provide your fig trees with the nutrients they need for vigorous growth and abundant fruiting. By honoring the earth's natural cycles and promoting biodiversity, you create an environment conducive to the long-term health of your fig trees and your ecosystem as a whole.

Chapter 161: Managing Specific Pests of Fig Tree: Protecting the Harvest Naturally

Growing fig trees offers a multitude of sweet delights, but it can also attract a variety of specific pests that threaten the health and productivity of these fruit trees.

1. Fruit Fly: A Greedy Invader

The fruit fly is one of the main pests of fig trees, damaging fruit by laying its eggs and forming galls. To combat them, pheromone traps can be used to capture male flies and disrupt their reproductive cycle.

2. Fig Aphid: Small Threat, Big Impact

Fig aphids feed on leaf sap, which can weaken trees. Introducing predatory insects like ladybugs and lacewings can help control their population naturally.

3. Defoliating Caterpillars: Targeted Biological Control

Some caterpillars feed on the leaves of fig trees, reducing their ability to photosynthesize. The introduction of natural parasites such as parasitoids and parasitic wasps can help contain their proliferation.

4. Mealybugs: Prevention and Control

Mealybugs, small sucking insects, can cause damage to fig trees by weakening their growth. Using soap or neem oil solutions can help control their presence.

5. Red Spiders: Combat at the Prudent Gardener

Spider mites feed on leaf sap and leave behind tiny webs. Regular watering of leaves and the introduction of natural predators like predatory mites can reduce their numbers.

6. Mites: Balancing the Ecosystem

Mites can feed on the leaves and stems of fig trees. Encouraging biodiversity in your garden, including habitat for predatory mites, can help keep their population under control.

7. Snails and Slugs: Physical and Biological Barriers

Snails and slugs can damage the fruits and leaves of fig trees. Using physical barriers like cups filled with beer can attract them and keep them away from trees. Natural predators such as ducks or ground beetles can also help control their population.

8. Use of Repellent Plants

Certain plants can act as natural pest repellents. Planting aromatic plants like mint or rosemary near fig trees can deter harmful insects.

The management of specific pests of the fig tree requires a balanced and environmentally friendly approach. By promoting biodiversity, using biological control methods, introducing natural predators and adopting appropriate cultural practices, it is possible to protect fig trees from pests without resorting to aggressive chemicals. Understanding pest life cycles and their interaction with the ecosystem is essential to preserving the health of fruit trees and ensuring abundant, healthy harvests.

Chapter 162: Fig Tree Cultivation in Subtropical Climate: Adaptation and Rewards

The fig tree, emblem of Mediterranean sweetness, can also thrive in subtropical climates, where temperatures are higher and ambient humidity is more pronounced.

1. Adaptation to Subtropical Climates

Fig trees, native to Mediterranean regions, can be successfully adapted to subtropical climates by following a few key principles. Choosing varieties adapted to higher temperatures and increased humidity is essential to ensure growing success.

2. Choice of Varieties

Some varieties of fig trees are better suited to subtropical climates than others. Varieties such as 'Black Mission', 'Brown Turkey' and 'Brown Turkey'. and 'Kadota' are known for their ability to tolerate higher temperatures and produce quality fruit in subtropical conditions.

3. Humidity Management

Subtropical climates can be characterized by periods of high humidity, which can encourage the development of fungal diseases. Air circulation, proper pruning to promote good ventilation and avoiding stagnant moisture around trees are key measures to reduce the risk of disease.

4. Adapted Irrigation

In subtropical climates, where rainfall can be erratic, regular and adequate irrigation is essential to ensure the growth and development of fig trees. It is important to maintain a balance between soil moisture and evaporation.

5. Protection Against Extreme Temperatures

Although fig trees can tolerate high temperatures, prolonged heat waves can affect their health. Providing partial shade during the hottest days and using mulch to maintain a stable soil temperature can help protect trees.

6. Pruning and Training Practices

Regular pruning of fig trees in subtropical climates is important to maintain a healthy structure and promote good air circulation. This helps reduce the risk of fungal diseases and maximize fruit production.

7. Balanced Fertilization

Fig trees in subtropical climates benefit from balanced fertilization to support their growth and development. The supply of essential nutrients, such as nitrogen, phosphorus and potassium, should be adjusted according to the specific needs of the variety and soil conditions.

8. Rewards of Culture in Subtropical Climate

Growing fig trees in subtropical climates can offer unique rewards. Warm conditions promote faster fruit ripening, creating a longer harvest season. Figs grown in these environments can also develop more intense aromas and flavors, particularly when exposed to temperature variations.

Growing fig trees in subtropical climates can be a rewarding experience for lovers of sweet, juicy fruits. By adapting cultivation practices to the specific conditions of these regions, it is possible to overcome challenges and harvest delicious fruits. Selection of suitable varieties, moisture management, protection from temperature extremes and regular maintenance are all factors that contribute to successful cultivation in a subtropical climate.

Chapter 163: The Fig and Ecological Solutions to Pest Problems: Towards Coexistence
Harmonious

Growing fig trees offers an abundance of sweet and nutritious delights, but it can sometimes be hampered by pests. However, instead of resorting to harsh chemicals, there are eco-friendly and sustainable solutions to prevent and manage pests in an environmentally friendly way.

1. Understanding Parasites

The first step in developing environmentally friendly solutions to pest problems is to understand the biology and behavior of specific pests that affect fig trees. This knowledge helps identify critical moments in their life cycle where preventative measures can be taken.

2. Use of Natural Predators

One of the most environmentally friendly approaches to controlling pests is to encourage natural predators. Ladybugs, lacewings and parasitoid wasps are examples of predators that feed on parasites such as aphids and caterpillars.

3. Crop Rotation

Crop rotation is an effective practice for reducing the buildup of pests in the soil. By changing the location of fig trees each year, pests that have a specific affinity for these trees will have more difficulty establishing themselves sustainably.

4. Use of Companion Plants

Certain plants act as natural repellents against specific pests. Planting aromatic herbs like mint, basil and sage near fig trees can discourage pests while adding an aromatic touch to the garden.

5. Pruning and Hygiene Practices

Maintain adequate pruning of fig trees and regularly remove damaged or infected parts can prevent the spread of parasites. Hygiene practices, such as collecting fallen leaves and disposing of them, also reduce potential breeding sites for pests.

6. Use of Natural Products

Natural insecticides made from substances such as neem, insecticidal soap, and baking soda can be used to treat pest infestations. These products are less toxic to the environment and non-target organisms than chemical pesticides.

7. Encourage Diversity

Creating a diverse ecosystem around fig trees can help balance the pest population. By providing a variety of plants, habitats and resources, you can attract a range of beneficial organisms that help control pest populations naturally.

Harmonious coexistence between fig trees and pests is possible thanks to ecological and sustainable solutions. Rather than compromising the health of the environment and fig trees with toxic chemicals, it is best to adopt ecosystem-friendly approaches. By understanding pest life cycles, using natural predators, promoting biodiversity and applying healthy growing practices, fig lovers can protect their harvests while helping to preserve the ecological balance.

Chapter 164: Cultivation of Fig Trees in Wind-Exposed Areas: Challenges and Solutions

Growing the fig tree can be rewarding, providing an abundance of delicious, sweet figs. However, in some regions fig trees face particular challenges, particularly areas exposed to wind.

Effects of Wind on Fig Trees
Wind-exposed areas can cause a variety of problems for developing fig trees. The most common effects include:

1. Dehydration: Wind can increase evaporation of moisture from leaves and soil, which can cause fig trees to dehydrate quickly.

2. Water Stress: The combination of wind and increased evaporation can cause water stress, making fig trees more vulnerable to diseases and pests.

3. Branch Breakage: Strong wind gusts can cause branches to break, which can damage the structure of the tree and compromise the future harvest.

4. Reduced Pollination: Wind can disrupt the pollination process, preventing fig flowers from pollinating properly, resulting in reduced fruit production.

Solutions for Growing Fig Trees in Windy Areas

1. Choice of Resistant Varieties: Opt for varieties of fig trees that are naturally resistant to strong winds. Some varieties feature tougher leaves and more flexible stems, making them better suited to windy conditions.

2. Wind Protection: Plant windbreak hedges, fences or other structures to protect fig trees from strong winds. This can help reduce wind force and create more sheltered areas.

3. Prudent Size: By practicing proper pruning, removing dead branches and maintaining a strong structure, you can reduce the risk of branches being broken by wind.

4. Adapted Irrigation: In windy areas, regular irrigation is essential to compensate for increased evaporation. Use drip irrigation systems or soakers to keep the soil moist.

5. Protection of Flowers: To protect flowers from wind disturbance, consider using netting or lightweight fabrics to cover them during the flowering period.

6. Solid anchoring: When planting fig trees, make sure they are properly anchored in the ground to withstand strong winds. A strong base and well-developed root structure are essential.

Growing fig trees in windy areas may seem difficult, but with the right practices and solutions it can be successful. By choosing hardy varieties, protecting fig trees from strong winds, maintaining a strong structure and providing adequate irrigation, gardeners can overcome wind challenges and harvest delicious figs. Perseverance and applying proper techniques are the keys to turning a challenge into an opportunity to grow healthy, productive fig trees, even in windy conditions.

Chapter 165: Fig and Organic Remedies Against Harmful Insects: Growing Healthily

Growing fig trees can be a rewarding experience, but as with any plant, it can be subject to attack from insect pests. Rather than resorting to harsh chemicals, many gardeners are adopting organic control methods to preserve the health of their fig trees and the environment.

Common Insect Pests for Fig Trees

Before exploring organic solutions, it is important to recognize some of the insect pests that can affect fig trees:

1. Aphids: These small insects feed on the sap of fig trees, causing the plant to weaken and the leaves to become distorted.

2. Thrips: Thrips feed on the leaves and fruit of fig trees, causing visible damage and potentially spreading viruses.

3. Cochineals: These pests attach themselves to the leaves and stems of fig trees, sucking up the sap and causing general weakening.

4. The caterpillars : Some caterpillars can feed on the leaves of fig trees, which can reduce the ability to photosynthesize and weaken the tree. Biological Solutions to Fight Insects

Pests

1. Prevention: The first line of defense is maintaining the overall health of the fig trees. A good floor balanced in nutrients, adequate irrigation and proper pruning can help build tree resistance against pests.

2. Use of Natural Predators: Introducing natural predators such as ladybugs, parasitoid wasps and titmice can help control pest insect populations.

3. Natural Repellents: Use natural repellents like neem oil, grapefruit seed oil or diluted garlic to discourage pests.

4. Pitfalls: Place yellow sticky traps or pheromone traps to attract and capture pests.

5. Insecticide soap: Use an organic insecticidal soap to gently eliminate pests. Make sure you don't affect beneficial predators.

6. Crop rotation: Crop rotation can help prevent insect pests from becoming permanently established in a given area. 7. Water Spray: Use a strong stream of water to physically remove insects from leaves and stems.

8. Natural Fertilizers: Use natural fertilizers rich in nitrogen to promote the healthy growth of fig trees, strengthening their resistance to pests.

Biological pest control on fig trees is an environmentally friendly approach that promotes the long-term health of the trees and the surrounding ecosystem. By using preventative methods, natural predators, repellents and gentle biological solutions, gardeners can maintain the vitality of their fig trees while minimizing negative impacts on the environment. Organic remedies make it possible to grow healthy and productive fig trees without the use of harmful chemicals, creating a harmonious balance between nature and agriculture.

Chapter 166: Protecting the Fig Tree from Extreme Weather: Strategies

to Preserve Vitality

Growing the fig tree can be a rewarding experience, but like any plant, it is subject to the vagaries of the weather, especially extreme weather conditions. Fig trees can be vulnerable to frost, extreme heat, strong winds and other weather events.

Preparing for Winter: Protecting against the Cold

1. Mulching: Applying a thick layer of mulch around the base of the fig tree helps protect the roots from intense cold and temperature fluctuations.

2. Wrap : Wrapping fig trees with materials like burlap or non-woven fabric during the winter can protect sensitive parts of the tree from frost.

3. Root Protection: Using an insulating material like straw to cover the roots during the winter can prevent excessive freezing. Adaptation to Intense Heat: Reducing Thermal Stress

1. Watering: During periods of intense heat, ensure regular and deep watering to maintain adequate humidity in the soil.

2. Shading: Use shade cloths or fabrics to create partial shade, helping to protect leaves and fruits from the hot sun.

3. Mulch: Apply a layer of organic mulch around the tree to preserve soil moisture and reduce evaporation.

4. Evening watering: Avoid watering during the hottest hours of the day. Prefer watering in the evening to minimize evaporation. Resistance to Violent Winds: Strengthening Structures

1. Tutors: Using stakes to support young trees can prevent them from lying down or breaking in strong winds.

2. Prudent Size: Removing dead or weak branches can prevent branches from breaking during high winds.

3. Windproof fence: Planting a hedge or windbreak fence can reduce the force of wind around fig trees.

Preventing Sudden Bad Weather: Being Proactive

1. Weather Monitoring: Monitor weather forecasts and take preventative measures based on upcoming conditions.

2. Quick Reaction: In the event of sudden extreme conditions, such as an unexpected frost in spring, respond quickly by applying protective methods.

3. Resistant Varieties: Opt for varieties of fig trees that are resistant to the specific weather conditions in your region.

Protecting the fig tree from extreme weather conditions is essential to maintaining its health and productivity. By adopting strategies for preparing for winter, adapting to intense heat, resisting violent winds and preventing sudden bad weather, gardeners can ensure the vitality of their fig trees in the face of climatic challenges. Regular weather observation and taking proactive preventative measures are key elements in preserving the resilience of fig trees and their capacity

à thrive despite changing climatic conditions. Through these efforts, fig trees will continue to provide delicious fruit and beautify our outdoor spaces, even in the most difficult of times.

Chapter 167: Fig Cultivation in Sandy Soils: Challenges and Solutions for Harvesting

Abundant

Growing fig trees in sandy soils can present both unique benefits and challenges. While sandy soils provide good drainage and high permeability, they tend to dry out quickly and lack essential nutrients.

The Benefits of Sandy Soils for Fig Trees

Sandy soils have some advantages for growing fig trees, including:

1. **Effective Drainage:** Sandy soils have exceptional drainage capacity, preventing water from accumulating around the roots and thus reducing the risk of rot.

2. **Early Heat:** Because of their ability to warm quickly, sandy soils encourage early growth of fig trees in spring.

3. **Improved ventilation:** The granular structure of sandy soils facilitates root aeration, which is beneficial to the health of the tree.

Sandy Soil Challenges for Fig Trees

However, sandy soils also present challenges:

1. **Limited Water Retention:** Sandy soils have low water holding capacity, which can cause water stress for fig trees, especially during hot, dry periods.

2. **Nutrient Depletion:** Nutrients are less likely to remain in sandy soils due to their high permeability, which can lead to nutritional deficiency for fig trees.

3. **Potential Erosion:** Due to their light structure, sandy soils are prone to erosion, which can damage roots and reduce tree stability.

Strategies for Growing Fig Trees in Sandy Soils

1. **Soil Improvement:** Adding organic matter like compost or manure can improve the water retention and fertility of sandy soil.

2. **Regular Irrigation:** Providing regular, deep watering is crucial to avoiding stress water of fig trees.

3. **Mulching:** Applying a thick mulch around the base of the fig tree can help conserve moisture, reduce erosion and provide nutrients to the soil.

4. **Balanced Fertilization:** Use balanced fertilizers to provide fig trees with essential nutrients

which they may lack in sandy soils.

5. **Choice of Suitable Varieties:** Opting for fig tree varieties that are more tolerant of sandy soils can increase the chances of success.

Growing fig trees in sandy soils requires special attention and proper care to ensure healthy growth and a bountiful harvest. By using water management, soil improvement and judicious fertilization practices, gardeners can overcome the challenges of sandy soils and enjoy the drainage and soil benefits they offer. early heat. With a proactive approach and careful care, it is entirely possible to grow lush, productive fig trees in sandy soil conditions, taking advantage of their unique qualities for a delicious and rewarding harvest.

Chapter 168: Using Natural Mulch to Promote Healthy Fig Roots

In fig tree cultivation, the well-being of the roots is essential to ensure vigorous growth and an abundant harvest. The use of natural mulches can play a crucial role in promoting root health by creating an environment conducive to their development and providing a range of benefits to the tree.

The Benefits of Natural Mulch for Fig Tree Roots

1. **Moisture Conservation:** Natural mulches, such as wood chips, fallen leaves or straw, retain soil moisture by creating a barrier that reduces evaporation. This keeps the roots of the fig tree hydrated, even during hot, dry periods.

2. **Temperature Regulation:** Mulches act as an insulating layer, protecting the roots of the fig tree from extreme temperature fluctuations. They minimize variations in heat and cold, providing a more stable environment for root growth.

3. **Erosion Prevention:** Mulches help prevent soil erosion caused by weather and wind,

thus keeping the roots of the fig tree firmly anchored and protected.

4. **Weed Control:** Natural mulches prevent weed growth, which reduces competition for nutrients and water and allows the fig tree's roots to thrive.

5. **Supply of Nutrients:** Some mulches, such as dead leaves and compost, gradually decompose and release nutrients into the soil, nourishing the fig tree's roots.

Mulching Practices for Fig Root Health

1. **Choosing the Right Mulch:** Opt for natural, non-chemically treated mulches, such as uncolored wood chips, straw, fallen leaves or walnut shells. Avoid mulch which could be harmful to the fig tree.

2. **Suitable Thickness:** Apply a layer of mulch about 5 to 10 centimeters thick around the base of the fig tree. Make sure you don't pile the mulch up against the trunk, which could cause rot.

3. **Regular renewal:** Mulches decompose over time, so it is recommended to renew them annually to maintain effective coverage.

4. **Ventilation:** Avoid compacting mulch, as adequate aeration is essential to allow roots to breathe.

5. **Distance from the Trunk:** Leave a space of about 10 centimeters around the trunk of the fig tree without mulch to prevent rot and inadequate air circulation.

Using natural mulches can significantly improve the health of fig tree roots by providing a range of benefits from moisture conservation to temperature regulation and weed suppression. By adopting proper mulching practices, gardeners can create an environment conducive to healthy root growth, thereby promoting growth and productivity global fig tree. The combination of natural mulch and proper care will help ensure that the roots of the fig tree thrive, leading to a bountiful harvest and a strong, thriving tree.

Chapter 169: The Advantages of Cultivating Fig Trees in Agroforestry

Agroforestry, a sustainable and integrated approach to agricultural land management, is growing in popularity due to its many ecological, economic and social benefits. Fig tree cultivation in agroforestry offers a unique opportunity to take advantage of the multiple benefits of this practice for ecosystems, farmers and local communities.

Improved Biodiversity

The integration of the fig tree into agroforestry systems contributes to the diversification of crops and the creation of habitats conducive to a variety of plant and animal species. Fig trees provide ecological niches for pollinating insects, birds and other beneficial organisms, promoting biodiversity and natural pest regulation.

Soil and Water Conservation

Agroforestry systems that include fig trees help reduce soil erosion by stabilizing land and minimizing runoff. The fig tree's deep root systems contribute to soil stability, while their plant cover reduces the impact of raindrops. Additionally, fig trees play a crucial role in water conservation by regulating the soil water cycle.

Nitrogen Fixation

Certain types of fig trees, such as strangler figs, are capable of fixing atmospheric nitrogen in the soil through a symbiosis with specific bacteria. This enriches the nutrient content of the soil, benefiting neighboring crops and reducing the need for chemical fertilizer applications.

Sustainable Food Production

Fig trees provide a diverse, nutrient-rich food source for farmers and local communities. Figs, rich in fiber, minerals and antioxidants, offer a healthy and tasty food option. Additionally, diversified agroforestry systems allow farmers to grow multiple crops simultaneously, thereby strengthening food security and income.

Improved Income and Livelihoods

Fig tree cultivation can provide an additional source of income for farmers. Fresh and dried figs can be sold in local markets or processed into value-added products such as jams, fruit jellies and cosmetics. This economic diversification strengthens the resilience of agricultural households in the face of market fluctuations.

Mitigating the Effects of Climate Change

Fig trees, as fast-growing trees, absorb carbon dioxide from the atmosphere, helping to mitigate the effects of climate change. Their presence in agroforestry systems helps regulate microclimate, reduce temperature extremes and improve crop resilience.

Growing the fig tree in agroforestry offers a multitude of benefits that go beyond simple fruit production. It promotes biodiversity, protects soil and water, enriches soil with nutrients, ensures sustainable food production and contributes to the fight against climate change. By thoughtfully integrating fig trees into agricultural systems, farmers can create balanced and resilient environments, promoting the long-term sustainability of ecosystems and farming communities.

Chapter 170: The Integrated Fight Against the Enemies of the Fig Tree

Integrated fig pest management, a holistic and sustainable approach, aims to maintain the health of fig trees by balancing management methods to minimize damage caused by pests, diseases and other pests. Rather than relying on harsh chemical solutions, this approach encourages the use of biological, cultural and physical methods to preserve fig trees and promote balanced ecosystems.

Principles of Integrated Pest Management

1. **Precise Identification:**The first step in integrated pest management is to accurately identify the specific pests and diseases that affect fig trees. This allows you to choose the best management methods adapted to each situation.

2. **Prevention:**Prevention is essential to avoid the introduction of pests and diseases. This may include measures such as using healthy, resilient plants, crop rotation and maintaining proper hygiene in orchards.

3. **Favor Natural Enemies:**Encouraging the presence of beneficial organisms such as natural predators, parasites and pollinating insects can help maintain ecological balance. For example, attracting ladybugs to fight aphids.

4. **Use of Physical Barriers:**Setting traps, nets or other physical barriers can prevent pests from reaching fig trees. This can be particularly effective in preventing actively moving pests.

5. **Cultural Methods:**Crop rotation, correct pruning, irrigation management and balanced nutrition can strengthen the resilience of fig trees and make them less vulnerable to pest attacks.

6. **Selective Use of Chemicals:**If necessary, the use of chemical pesticides should be the last option and should be carried out selectively to minimize negative impacts on the environment and human health.

Advantages of Integrated Pest Management

1. **Environmental Sustainability:**By favoring biological and non-chemical methods, integrated pest management respects biodiversity and preserves natural ecosystems

2. **Human health :**Reducing the use of chemical pesticides decreases exposure to toxic products, protecting the health of farmers and consumers.

3.**Profitability:** Integrated pest management can reduce the costs associated with the purchase of pesticides, while promoting better long-term productivity.

4. **Preservation of Fruit Quality:**By minimizing pest and disease attacks, integrated pest management contributes to the production of better quality fruit.

5. **Long-Term Resistance:**Adopting sustainable methods strengthens the resilience of fig trees, thus preventing the development of resistance to pesticides.

Integrated Fig Pest Management is an effective, environmentally friendly approach to maintaining tree health while minimizing the use of toxic chemicals. By combining knowledge of pests and diseases, judicious cultural practices and organic methods, this approach promotes more balanced ecosystems, better fruit quality and long-term sustainable agricultural systems.

Chapter 171: Fig Tree Cultivation in Limestone Soils: Challenges and Solutions

Growing fig trees in limestone soils presents a unique challenge for farmers and gardeners. Limestone soils, rich in calcium carbonate, can have an impact on the nutrition and growth of fig trees. However, with suitable practices, it is possible to overcome these challenges and successfully grow productive and healthy fig trees under such conditions.

The Challenges of Limestone Soils

1.**Nutrient Needs:**Calcareous soils can limit the absorption of certain essential nutrients, such as iron, zinc and copper, due to the fixation of these elements by calcium.

2. **Soil reaction:**Limestone soils tend to be alkaline, which can affect the uptake of nutrients by the fig tree's roots, particularly elements such as iron, manganese and zinc.

3. **Compaction and Drainage:**Limestone soils can sometimes be compacted, which limits drainage and can lead to water stagnation, causing root rot problems.

Solutions for Growing Fig Trees in Limestone Soils

1. **Choice of Varieties:**Some varieties of fig trees are more tolerant of limestone soils than others. It is important to choose varieties adapted to these conditions to increase the chances of success.

2. **Organic Amendments:**Adding organic matter in the form of compost or manure improves the structure of calcareous soils, improves drainage and provides additional nutrients.

3. **Targeted Nutrient Intakes:**Depending on the specific nutrient needs of the fig tree, it may be necessary to apply fertilizers rich in micronutrients such as iron, zinc and manganese. However, this should be done in moderation to avoid imbalances.

4. **Nutrient Chelation:**Using chelated amendments can help make certain nutrients more accessible to roots by preventing them from being uptake by calcium.

5. **Watering Management:**Adequate drainage is essential in calcareous soils. Avoiding over-irrigation and accumulation of water around the roots will prevent rot problems.

6. **Application of Lime:**In some cases the application of lime may be necessary to adjust the pH of the soil, although this should be done with caution and according to actual needs.

Advantages of Growing Fig Trees in Limestone Soils

1. **Disease Resistance:**Limestone soils can sometimes be less favorable to the development of

certain diseases, which can contribute to the overall health of fig trees.

2. **Resistance to Salinity:** In some areas, calcareous soils can be associated with high levels of salinity. Fig trees tend to tolerate these conditions better than other crops.

3. **Improved Flavors:** Fig trees grown in calcareous soils can produce fruits with flavors and more concentrated aromas.

Although growing fig trees in calcareous soils presents specific challenges, these challenges can be overcome with suitable management practices. By choosing the right varieties, amending the soil with organic matter and providing essential nutrients in a targeted manner, it is possible to grow productive and healthy fig trees under these conditions. With careful attention to irrigation and root environment management, gardeners and farmers can enjoy the benefits of growing fig trees in calcareous soils

Chapter 172: Fig and Disease Prevention Methods: Healthy Cultivation

The fig, with its rich history and succulent taste, is a treasure for fruit lovers. However, for fig trees to thrive and produce abundant fruit, it is essential to protect them from diseases that can weaken them. Fig tree disease prevention relies on carefully planned growing practices and appropriate management strategies.

1. Choice of Resistant Varieties: Opting for varieties of fig trees known for their disease resistance is a first line of defense. Some varieties naturally have better resistance to certain infections.

2. Healthy Soil and Drainage: Well-drained, healthy soil is crucial to preventing moisture buildup that encourages the development of fungal diseases. Good soil structure also promotes vigorous roots and healthy growth.

3. Avoid Excess Humidity: Watering the fig tree adequately is essential. Avoid watering excessive which can cause the development of mold and fungus.

4. Tool Hygiene: Using clean, disinfected tools when pruning or other work minimizes the spread of pathogens.

5. Correct Size: Proper pruning allows adequate air circulation within the tree, preventing moisture from becoming trapped and reducing the risk of disease.

6. Removal of Diseased Leaves: Remove infected leaves regularly to prevent the spread of disease.

7. Use of Mulch: Applying organic mulch around the base of the fig tree helps keep the soil clean and prevents splashes of contaminated water from hitting the leaves.

8. Crop rotation: If possible, avoid planting fig trees in the same location every year. Crop rotation prevents the accumulation of pathogens in the soil.

9. Natural Treatment: Some natural herbal or mineral solutions can help prevent disease. For example, a spray of diluted baking soda can help prevent fungal infections.

10. Regular Monitoring: Inspect your fig trees regularly to quickly identify any signs of disease. Early intervention is often the key to preventing spread.

11. Biological Resistance: Improving the overall health of fig trees, through the use of proper cultural practices, can enhance their natural resistance to disease.

12. Balanced Fertilization: Providing your fig trees with a balanced diet of nutrients promotes healthy growth and greater disease resistance.

13. Protection Against Pests: Certain pests can damage fig trees and make them more susceptible to disease. By controlling pests, you indirectly reduce the risk of diseases. Ultimately, a combination of preventative practices is the key to growing healthy fig trees. Vigilance, knowing the warning signs of disease and adopting environmentally friendly methods are effective ways to preserve the health of your fig trees and harvest delicious and abundant fruit.

Chapter 173: The Importance of Adequate Irrigation for the Fig Tree: Nourishing Prosperity

Irrigation, or water supply, is a fundamental component of growing any tree, including the fig tree. This crucial practice determines the health, growth and productivity of the tree. The fig tree, with its thousand-year-old history and its taste appeal, deserves special attention in terms of irrigation.

1. Providing Necessary Water: The fig tree needs an adequate amount of water to grow properly. Regular irrigation ensures healthy growth of roots and aerial parts, thus promoting fruit production.

2. Meeting Variable Needs: The water requirements of the fig tree vary depending on factors such as season, temperature, growth stage and size of the tree. Irrigation should be adjusted accordingly to respond to these fluctuations.

3. Avoid Excess Water: Excess water can cause root suffocation and promote the development of fungal diseases. Excessive irrigation should be avoided to maintain optimal balance.

4. Encourage Fruiting: Regular irrigation during the fruit development period encourages fig growth and abundant production of quality fruit.

5. Prevention of Water Stress:A lack of water can stress the fig tree, leading to slowed growth, wilted leaves and reduced fruit production.

6. Strengthen Resistance: Proper irrigation helps strengthen the fig tree's resistance to diseases and pests while maintaining its overall health.

7. Water Saving: Well-planned and managed irrigation allows water to be used efficiently, avoiding waste while still meeting the needs of the tree.

8. Irrigation Methods: Different irrigation methods, such as drip, porous hose or hand watering, can be used depending on local conditions and grower preferences.

9. Avoid Splash: Direct watering on the leaves can encourage the development of diseases. It is best to water the soil directly to minimize splashing.

10. Consider the Type of Soil: Clay soils retain water longer than sandy soils. The frequency and duration of irrigation should be adjusted depending on the soil type.

11. Avoid Surface Irrigation: Surface watering can encourage root growth near the surface, making them vulnerable to extreme weather conditions.

12. Use Technology: Smart irrigation systems can help monitor and control water supply based on the needs of the tree.

In short, adequate irrigation is an essential element to guarantee the prosperity of the fig tree. Careful attention to water requirements, seasonal variations and the overall health of the tree can lead to vigorous growth, abundant fruiting and the preservation of figs' distinctive beauty and flavor. A careful balance between water intake and other cultural care forms the foundation for success in growing this tree valued since ancient times.

Chapter 174: Fig Tree Cultivation in Frost Zones: Challenges and Rewards

Growing fig trees in areas of frost presents an exciting challenge for gardeners who want to grow this iconic fruit tree despite less favorable climatic conditions. If the fig tree is often associated with warm Mediterranean regions, it is possible to successfully cultivate it in areas frost-prone, using specific techniques and understanding the unique needs of the tree.

1. Choice of Resistant Varieties: The first step to successfully growing fig trees in frost zones is to choose cold-resistant varieties. Some varieties, such as 'Chicago Hardy', 'Brown Turkey' and 'Brown Turkey'. and 'Celeste', are more adapted to cold climates and tolerate lower winter temperatures.

2. Preparation for Cold: Before winter arrives, it is recommended to prepare the fig tree by watering it abundantly and mulching the ground around the tree. This helps protect the roots from freezing.

3. Strategic Location: Plant the fig tree in a location that gets full sun, ideally against a wall or fence which can act as a natural thermal radiator and help retain heat.

4. Use of Winter Sails: Cover the fig tree with overwintering sails before frost periods to protect the branches and buds from intense cold.

5. Root Protection: Protect the roots of the fig tree by applying a layer of thick mulch around the base of the tree. This helps maintain a more stable soil temperature.

6. Reflected Size: Pruning the fig tree before winter can help reduce frost damage by removing weak parts and encouraging the growth of strong new branches.

7. Grafting and Wrapping the Stems: Some gardeners choose to graft cold-hardy varieties onto rootstocks more suited to winter conditions. Wrapping the stems with insulating material, such as straw, can also help minimize damage.

8. Patience and Observations: As temperatures begin to rise in spring, keep a close eye on buds and new growth. If damage has occurred, it may take time for the fig tree to recover. Be patient and provide necessary care.

9. Unique rewards: Growing fig trees in frost zones may seem intimidating, but the rewards are manifold. Harvesting fresh figs in a climate where they are not naturally abundant can bring great satisfaction. Figs grown in freezing areas can often be sweeter and more flavorful due to temperature variations that influence the concentration of sugars in the fruit.

10. Connection to Nature: Growing the fig tree in difficult conditions strengthens the connection with nature and allows gardeners to adapt and learn to work in harmony with their environment.

Growing fig trees in frost zones requires special attention and careful preparation, but it can be rewarding for determined gardeners. With the right techniques and an understanding of the fig tree's specific needs in cold conditions, gardeners can enjoy the delights of this exceptional fruit even in cooler climates.

Chapter 175: Fig and Pest Control Practices: Balancing Protection and **durability**

Fig growing is often an exciting and rewarding endeavor, but it is not without its challenges, including the presence of insect pests that can damage the leaves, buds and fruit. Managing insect pests in fig cultivation is a crucial aspect to ensure healthy and abundant yields while maintaining ecosystem sustainability.

1. Precise Identification: The first step to managing insect pests in fig growing is to correctly identify the specific pests present in the area. Each species may require different control strategies.

2. Prevention Methods: Prevention is the key to avoiding a major infestation. Use prevention methods such as planting insect-resistant varieties, keeping the garden clean and hygienic, and using nets or sails to prevent insects from landing on plants. fig trees.

3. Cultural Practices: Maintain the vigor of fig trees through healthy cultural practices, such as

Regular pruning, proper watering and adequate fertilization, can enhance their natural resistance to harmful insects.

4. Use of Natural Predators: Encouraging the presence of natural predators, such as ladybugs and parasitoid wasps, can help control pest populations.

5. Using Traps: Using sticky traps or specific pheromonal traps can help capture certain pests and reduce their numbers.

6. Organic Insecticides: In the event of an infestation, favor biological insecticides based on bacteria, fungi or other natural agents specific to pests.

7. Crop rotation: Practicing crop rotation helps disrupt the life cycle of insect pests, thereby reducing their impact on fig trees.

8. Avoid Excessive Treatments: Excessive use of insecticides can have detrimental effects on the ecosystem by eliminating natural predators and causing resistance in targeted insects. Use treatments targeted and sparingly.

9. Monitor Regularly: Regular monitoring of fig trees is essential to quickly detect any incipient infestations and take action before pest populations get out of control.

10. Education and Awareness: Raise awareness of environmentally friendly pest control methods and encourage other gardeners to adopt sustainable practices.

It is important to understand that excessive use of chemicals can have harmful consequences on human health, pollinators and the environment in general. Therefore, implementing environmentally friendly pest control practices is essential to maintain balance in the garden and preserve the health of the fig trees as well as the surrounding biodiversity. Ultimately, the combination of preventative methods, biological techniques and management responsible for harmful insects will contribute to a prosperous and sustainable fig culture.

Chapter 176: Selecting Fig Varieties Suitable for Your Area: Cultivating Successfully

Successfully growing fig trees depends largely on choosing the right varieties for your specific region. Each region has its own unique climate, soil and growing conditions, and selecting the appropriate fig tree varieties can ensure abundant, high-quality yields.

1. Climate : The climate of your area is one of the most important factors to consider when choosing fig tree varieties. Some varieties are better suited to hot, dry climates, while others can withstand colder winters. Be sure to choose varieties that are compatible with the temperatures and weather conditions in your area.

2. Length of Growing Season: Some varieties of fig trees require a long growing season to produce ripe fruit. If you live in an area with short summers, opt for earlier maturing varieties to ensure a successful harvest.

3. Ground : The type of soil in your area influences the growth and development of fig trees. Some soils are more draining, while others retain more water. Choose varieties that adapt well to the soil in your area.

4. Disease Resistance: Some varieties of fig trees are more resistant to common diseases such as downy mildew or rust. Opting for resistant varieties can reduce the need for chemical treatments.

5. Tree Size: The mature size of the tree is also a factor to consider. If you have limited space, choose dwarf or semi-dwarf varieties that will best suit your garden.

6. Flavor and Use: Fig trees produce fruit with a range of flavors, textures and colors. Choose varieties that match your taste preferences and how you want to use them, whether for fresh consumption, cooking, jamming or drying.

7. Resources Available: Consider the resources available in your area, such as water supply and care requirements for the varieties you are considering. Some varieties may need more attention than others.

8. Local Research and Advice: Research varieties that have already been grown successfully in your area. Advice from local gardeners, nurserymen or horticultural experts can also be valuable in helping you choose the most suitable varieties.

9. Experimentation : It may be a good idea to experiment by planting several varieties to determine which ones thrive best in your area. Observe growth, fruit production and resistance to local conditions.

Choosing fig tree varieties suited to your region is an essential step in ensuring the success of your crop. By carefully considering climate, soil, disease resistance and tree pruning factors, you can create a thriving and productive fig garden that thrives harmoniously in your local environment.

Chapter 177: Fig Cultivation in Acidic Soils: Challenges and Solutions

Growing fig trees in acidic soils can present unique challenges, but with careful planning and proper care, it is possible to grow these delicious fruits in less pH-friendly conditions.

Challenges :

1. **Soil pH:**Fig trees generally prefer neutral to slightly alkaline soil, with a pH between 6 and 7. Acidic soils, having a pH below 6, can make it difficult for fig tree roots to absorb essential nutrients.

2. **Inaccessible Nutrients:**In acidic soils, some vital nutrients such as calcium, magnesium and phosphorus can be chemically bound and become less available to plants,

which can lead to nutritional deficiencies.

The solutions :

1. **Soil Test:** Before planting fig trees in acidic soil, it is recommended to have the soil tested to determine its exact pH. This will help understand how acidic the soil is and what adjustments may be necessary.

2. **Limestone Amendments:** To raise the pH of acidic soil, adding limestone amendments such as agricultural lime can be effective. This will help make the soil more neutral or slightly alkaline, which is best suited for fig trees.

3. **Nutrient Enrichment:** Acidic soils may lack certain essential nutrients. By adding nutrient-rich organic amendments such as compost, well-rotted manure or slow-release fertilizers, you can improve soil quality and provide fig trees with necessary nutrients.

4. **Organic Mulch:** Using organic mulch around the base of fig trees can help maintain soil moisture and create an environment conducive to growth, also promoting the decomposition of organic matter which can help gradually adjust soil pH.

5. **Choice of Varieties:** Some varieties of fig trees are more tolerant of acidic soils than others. Do your research to choose varieties that have shown better adaptation to acidic conditions.

6. **Monitoring and Adjustments:** Once you have the amendments and recommended practices in place, regularly monitor soil pH and fig tree health. If necessary, make additional adjustments to maintain a supportive environment.

7. **Regular Irrigation:** In acidic soils, regular irrigation is important to maintain constant humidity. Make sure you don't let the soil dry out excessively, which could increase acidic pH challenges.

Growing fig trees in acidic soils may require a little more care and effort to create an environment conducive to their growth and fruit production. By following practices such as soil testing, amendment, nutrient enrichment, and choosing appropriate varieties, gardeners can overcome the challenges of acidic soil and enjoy a successful harvest delicious figs.

In conclusion, "Fig Trees Galore: The Global Encyclopedia" took us on a captivating journey through the rich and diverse world of fig trees. From the quest for the perfect variety to propagation, from traditional cultivation to innovative methods, from spiritual rituals to medicinal uses, this book explored every facet of this amazing tree that has shaped cultures and cultures.

civilizations through the ages.

À Through these pages, we have discovered how fig trees have transcended geographic and cultural boundaries, leaving their mark in the arts, literature, gastronomy, spirituality and much more. From ancient mythology to modern sustainable growing practices, each chapter of this book has shown us how deeply ingrained fig trees are in our lives.

Throughout this encyclopedia, we have explored the countless uses of fig trees, from traditional recipes to medicinal applications to their crucial role in biodiversity and environmental preservation. We have witnessed how fig trees have inspired art, spirituality, cuisine and technology throughout the ages, while remaining a symbol of sustainability, connection and growth.

"Fig Trees Galore: The Global Encyclopedia" embodies the passion and devotion of fig tree lovers around the world. Through these pages, we have not only expanded our knowledge of these remarkable trees, but we have also discovered the incredible potential they offer for the future of our planet and our societies.

As we close this book, we are invited to continue our journey with the fig trees, to plant and cherish them in our own lives. Whether for their succulent fruits, their rich symbolism spiritual or their ability to nourish our planet, fig trees remain a powerful reminder of our connection to nature and our role as stewards of the earth. "Fig Trees Galore: The Global Encyclopedia" will remain a valuable resource for anyone who wishes to thoroughly explore the life, history, and many dimensions of fig trees. May this book continue to inspire and enlighten, just as the fig trees themselves continue to enrich our lives with their fruit, their beauty and their beneficial presence.